Automation
Network
Selection:
A Reference Manual
2nd Edition

Automation Network Selection:
A Reference Manual
2nd Edition

by Dick Caro

Notice

The information presented in this publication is for the general education of the reader. Because neither the author nor the publisher has any control over the use of the information by the reader, both the author and the publisher disclaim any and all liability of any kind arising out of such use. The reader is expected to exercise sound professional judgment in using any of the information presented in a particular application.

Additionally, neither the author nor the publisher have investigated or considered the effect of any patents on the ability of the reader to use any of the information in a particular application. The reader is responsible for reviewing any possible patents that may affect any particular use of the information presented.

Any references to commercial products in the work are cited as examples only. Neither the author nor the publisher endorses any referenced commercial product. Any trademarks or tradenames referenced belong to the respective owner of the mark or name. Neither the author nor the publisher makes any representation regarding the availability of any referenced commercial product at any time. The manufacturer's instructions on use of any commercial product must be followed at all times, even if in conflict with the information in this publication.

Copyright © 2009

ISA–International Society of Automation
67 Alexander Drive
P.O. Box 12277
Research Triangle Park, NC 27709

Printed in the United States of America.
10 9 8 7 6 5 4 3 2

ISBN-13: 978-1-934394-89-2

Library of Congress Cataloging-in-Publication Data (in process)

Acknowledgments and Dedication

Working with a communications standards committee such as the ISA/IEC Fieldbus committee has been an honor and a privilege. Participating with some of the most brilliant engineers in the world has given me a perspective and education into communications protocols not possible in any other way. I want to thank these people and their employers, too many to name here, for allowing me the opportunity to learn in this way. Working with these highly intelligent and personable people from more than 17 countries has given me an international perspective that would not be possible to obtain in any other circumstances. I would also like to thank my past employers for giving me the opportunity to participate in the work of Fieldbus, but most of them are no longer in existence (Autech Data Systems, Computer Products, and Arthur D. Little).

I truly appreciate the help provided by many of the open bus trade organizations by making their works available through the Internet, or allowing me to read documents otherwise not available without charge.

As a communications author I wish to especially thank Tim Berners-Lee (whom I have never met), the acknowledged inventor of the Internet, without which this book and most of my work for the past decade or more would have not been possible. Likewise, I would like to thank the late Don Estridge of IBM (whom I did meet) for his leadership in creating the open market for the personal computer, without which works like this could take too long to bother.

Finally, I would like to thank Andy Chatha and all of my friends at ARC Advisory Group who gave me an international bully pulpit for several years and supported my work to complete the Fieldbus standard.

I would like to dedicate this book to Dr. Ted Williams, formerly Director of the Purdue Laboratory for Applied Industrial Control (PLAIC), Organizer and Chairman of the International Purdue Workshop on Industrial Computer Systems, President of ISA, and a good friend for many years. Ted's vision and persistence have seeded many standards for industrial automation, including ANSI/ISA5, 50, 61, 84, 88, and 95.

Table of Contents

Unit 1:
Prologue

Why are there so many different industrial automation networks? Why can't there be just one network? We hear these questions very often. Indeed, it could have been, but... Well, that would be a long story.

Here is the short version. In 1985, a bunch of us recognized the need for standardizing network communications for process control and we tried to prepare a standard in advance of the competition under the ISA50 standards committee. We called our effort *fieldbus* since it was meant for field instruments. Almost immediately, we were joined by suppliers of programmable logic controllers (PLCs) who believed that they needed to standardize upon their remote I/O networks and could share the same technology. We completed the physical layer protocol (wiring/cabling and signaling) in 1989 and the data link layer protocol in 1993. However, by then there were already numerous competitive commercial networks. The standards work was eventually completed in 1999 with the adoption of many of these commercial network architectures into the fieldbus standard IEC 61158 (International Electrotechnical Committee), which initially contained eight different protocols (types) and now contains more than 20 types.

Market forces could not wait for a standard, and the standard could not incorporate all market forces into a single protocol. Not only that, but a standard cannot immediately, if ever, displace already implemented commercial products. Seven of the original eight network technologies included in the IEC Fieldbus standard have been implemented in commercial products including the two specifications prepared by the Fieldbus Foundation that were based on IEC 61158 Types 1 and 5 that were originally developed from the ISA50.02 Fieldbus standard. One protocol has never been implemented.

I chaired both ISA50 and the IEC SC65C/WG6 Fieldbus committees during the end of their efforts. They were very turbulent times. Cullen Langford, a user from DuPont, chaired these committees during most of the time that they were creating the fieldbus protocol, while I was a contributing member of the user layer subcommittee. Being surrounded by some of the brightest people in the universe was a rare privilege and the experience of a lifetime. I didn't enjoy many of the political moves being made by major manufacturers, but it did provide a challenge to my management style.

While politics were raging in the ISA/IEC committees, discrete automation identified by the use of PLCs was also developing rapidly. For a while, it appeared that one of the standards bodies, NEMA (National Electrical Manufacturers Association), was going to use the high-speed version (H2) of ISA's fieldbus. However, NEMA concluded that H2 standards were too expensive and too complex for use in discrete manufacturing. As a result, most of the data communications buses developed by PLC manufacturers were incompatible with each other and did not conform to a standard set of specifications. Most of the manufacturers countered criticism about the "closed" nature of these protocols by forming "open" groups to make the long-term evolution for each of these bus technologies independent from the originating manufacturer. These open bus associations contributed five of the additional protocols to the IEC eight-part fieldbus standard.

In retrospect, it is clear that no single bus technology can satisfy the demands for multiple applications in the manufacturing marketplace. This is a political, not a technological, statement. We always knew that the wiring, cabling, and connecting solutions embedded in the physical layer could not span all markets, but the committees used the technique known as the "meld of best features" method of standardization to combine the simple elegance of WorldFIP with the pragmatism of Profibus. In the process, we succeeded in developing a single protocol meeting the needs of both process control and factory automation, but with a complexity that caused it to not be accepted for discrete manufacturing. It was over these issues that the final approval of the IEC 61158 fieldbus standard was delayed for seven years.

Also dating from the late 1980s were the development of very simple buses for sensors. These, too, were developed by independent manufacturers and typically share nothing in common with each other or with any of the higher level bus architectures. They were designed for low cost and low complexity. However, this did not prevent some of these bus structures from being promoted by their sponsors for higher level applications, thereby adding to the confusion.

This book gives you a perspective on the typical applications for industrial automation bus technology. The emphasis is upon the intended application for each bus, rather than the range of applications for each bus, which you would find in the supplier's literature. With that goes a note of caution: Any bus can generally be used for any application; however, stretching a bus technology outside its intended area creates more problems than it solves.

We will begin by discussing some bus applications and will propose the bus technologies that should be used to provide the needed communications services. Often, several different bus solutions will be appropriate, and their differences in the application context will be discussed. Then we will discuss the bus technologies and requirements from an end user point of view. Although the bus technology sections will talk briefly about the protocol used on the bus, the emphasis is on the wiring/cabling issues and the user interfaces. This book is generally free of mathematics except in those areas where the numbers have real application relevance.

Finally, I have always maintained my independence from all manufacturers to retain an unbiased viewpoint. This is an unbiased book, except for my fondness for the fine points of the fieldbus standard. I continue to be amazed at how little this formalized body of knowledge is used in industry, and I do reference some of these points.

Unit 2:
How to Use this
Book

There are many books about networks, some written as implementer guides, some written as historical records of a protocol development, and some written as textbooks for the network protocol designer. This is different! This book is directed at the manufacturing process designer to help select the "right" network to use for a given application. I would like to have called it *Industrial Networks for Dummies* but that title is too similar to already existing books for IT (Information Technology) networks. The book is divided into two major sections: a discussion of applications and a reference guide for the various networks. Although there are some applications for which only one network protocol is the preferred answer, most applications can be served by more than one. In some cases, the answer is none of the above, but that only gives opportunity for more than one network architecture to be used.

2.1   Application Guide

This portion of the book describes many classical network applications and the way in which these networks are used. I hope that you will find your application in this section, but I cannot possibly describe all applications. So, your job is to find an application most like yours. Very often, the problem is that your new networking project will have more than one application within it and they may be very different from each other. The normal answer is to use two or more different network architectures, but that may cause a measurable increase in installation cost and an even larger increase in long-term support/maintenance costs.

Network architecture selection is still an art form, not a science. Most of the network architectures have considerable overlap, so that many applications can be done with several networks. Pragmatically, most of the network architectures originated with the work of a single system or product manufacturer and are optimized about those applications with which they were most familiar. For this reason, it may be necessary to compromise your network selection to accommodate the network supported by your major system or equipment manufacturer.

2.2   Technology Guide

Someone has identified more than 29 industrial networks currently offered for sale. Many of these networks are proprietary to a single manufacturer and have a broad installed base. Sometimes, "friends" of these manufacturers are permitted to offer compatible, but not competitive, equipment using the same proprietary network architecture. The trend throughout industrial automation is to move away from proprietary networks and toward open or standard network architectures. For this reason, this book will not review proprietary network architectures. If the reader is determined to use these networks, they should contact the original supplier.

Network architectures are discussed in terms of the following criteria:

- Physical wiring/cabling requirements, sometimes called the *wiring plant*

- Power and safety requirements, including grounding

- Protocol in terms of efficiency, latency, and determinism

- Application support for the end user

Industrial automation networks are divided by class into sensor networks, remote I/O, smart I/O instrument networks, and control networks. Some network architectures overlap these classifications, and each variation is presented together with its peer networks for direct comparison. This is difficult to explain, but you will know it when you see it. Overlap comes from attempts by the network designers to move the technology into as many niche applications as possible, sometimes with great success, and many times with failure.

Unit 3:
Introduction to
Industrial Networks

At one time, it was thought that industrial automation networks were different from the kinds of networks used for IT (information technology.) In fact, the earliest automation networks were not even considered as networks at all but as serial buses. The term *fieldbus* stems from these thoughts. Naturally, each network was designed to solve one problem, then extended to solve other, perhaps related, problems. Since each supplier's business model was directed toward a slightly different business niche, the resulting *bus* turned out to be different from any other.

As long as industrial automation networks were slow and uncomplicated, no special components were required. For example, EIA[1]-232, EIA-422/423, and EIA-485 were often used for the physical layers, supported by commodity semiconductors. Early protocols were simple enough to execute on 8-bit microprocessors such as the 8051, Z80, or 6809. When speeds became higher and protocols richer in functionality, custom silicon became necessary to implement these networks. Custom silicon is expensive to design (nonrecurring engineering or NRE) and because the volumes are small compared with volumes in the IT market, expensive to manufacture.

The first approach to fix this problem was to standardize industrial automation networks through standards committees or the establishment of defacto standards by opening the specifications to multi-vendor committees. The theory was that if

---

1.   Some people remember this as RS-232, but the standards organizations, Electronics Industry Association and the TIA or Telecommunications Industry Association, prefer to designate it as EIA or TIA-232.

many system suppliers used the same chip set, there would be an economy of scale and a lower manufacturing cost. It just didn't work! There were too many chip sources and independent chip designs. The NIH (Not Invented Here) syndrome prevailed.

The trend is clearly to shelve the idea that industrial automation networks are somehow different from IT networks. The clear trend is to use COTS (commercial off-the-shelf) components and adapt them through software to industrial automation applications. Since Ethernet was the clear winner in the IT market, it is no surprise that Ethernet is the basis for the newest evolution of industrial automation networks—at least at the high-performance end—which leaves the lowest-level networks used for connecting sensors and actuators with a different solution. These networks are also migrating and converging in both directions with commodity silicon as the basis. Some low-level networks will most likely use scaled-down versions of the higher-performance industrial automation networks, whereas others will use low-cost silicon developed for other markets.

Finally, it should be obvious by now that any discussion of industrial automation networks must consider the software used for the upper layer visible to the end user. All network architectures are described by the ISO (International Standards Organization) standard OSI (Open Systems Interconnection) basic reference model: standard ISO/IEC 7498-1:1994. This model is illustrated in Figure 3-1 and is divided into seven parts. When we say *network protocol*, we are talking about stuff in these layers. The end user only cares about the connection to the physical wires coming out the bottom and the features and functions made available at the top. Yes, the protocol is important, but a lot less important than the claims made by the network designers.

Notice that there are two layers above the ISO/OSI seven layers. The OPC (object linking and embedding for process control) layer has the benefit of adapting the network layers to the host system. Thus the client user layer only needs to be created knowing that it will be used with a server running compatible

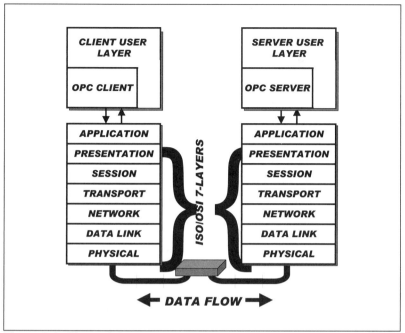

Figure 3-1. Network Layers

OPC software. With OPC, the details of the network layers are effectively hidden from view. It should also be noted that there are other methods of isolating the network application layer from the user layer software by using other network technologies, incorporated in the user layer, that do not use OPC. This is illustrated by the direct coupling of the user layer to the top of the communications protocol stack. Usually this is done to take advantage of the efficiency of the user layer connections and to make data transfers more deterministic than allowed by OPC.

Finally, in Figure 3-1, the network wiring is drawn with a little box between client and server. This box represents the physical network as more than just wire and cable. Very often, there are active switches and converters in these networks for a variety of reasons that we will discuss later when each network is described. The word *cable* is often used since it includes metallic wiring and fiber optics. It also should include wireless connections, but that would require the use of the oxymoron *wireless cable*, an expression actually in use in the telecommuni-cations market but not yet in industrial automation.

Next, we will introduce the generic types of networks used in industrial automation.

3.1   Sensor Networks

At the very lowest level of network functionality are the sensor networks. Generally, sensors themselves are at the bottom of the industrial automation system food chain and are designed to be inexpensive, since many sensors (and actuators as well) are required. Sensors provide basic data to the control system, such as the position of an object or a physical property such as temperature. Their mechanisms vary according to the desired accuracy and reliability.

The simplest sensor is typically the mechanical limit switch used to indicate that an object is present or not. For example, limit switches often are used to detect boxes on a conveyor belt as it passes a reading station. They are also used on control valves to indicate when the valve is fully open or fully closed. Limit switches must be powered (sometimes called *biased*) to detect the open/closed switch position. When used with a PLC, each limit switch typically requires two wires to connect to the I/O terminations on a digital input card/module inter-face that is plugged into a slot on a multiplexer (most often called a remote I/O or block I/O unit). These digital values are typically reported to the controller as a bit in an input register.

Other technologies used to detect position are photocells and proximity detectors. Both of these technologies are somewhat more expensive than limit switches, but provide no more infor-mation than does the ON/OFF switch position. However, they do not have moving parts, and generally have a longer service life. Photocells require a source of light in addition to the light-sensitive detector. Proximity detectors may be magnetic, and require no separate power source to detect iron or steel (mag-netic) objects. When the object is composed of non-magnetic materials such as paper, plastic, or aluminum, a power source is used to generate an inductive field that will be modified by the mass of the object and can be detected.

Sensor networks are designed to reduce the point-to-point wiring needed to connect the limit switch, proximity sensor, solenoid valve, or photocell to the I/O interface. This is done in two ways: (1) put a network driver inside the sensor or actuator itself, or (2) bring the I/O interface close to the sensor or actuator so that connections are very short. There are products on the market that do it both ways. The I/O interface usually terminates 4 to 16 I/O points and is connected to the PLC or other type of controller by the sensor network that transmits digital data for all points.

Sensor networks work by actually detecting the status of the sensor and converting it to a 1 or 0 in a status word. The status word is then transmitted across the network to a terminating device called a *scanner* that is usually in a remote I/O rack, a PLC, or a computer. The scanner is responsible for assembling the status words from each sensor network node into a register in the device. Each sensor network has its own method for mapping the sensor status to the I/O registers. The distinguishing factor of sensor networks is that the sensor, actuator, and network node do nothing more than convert the sensor or actuator state to or from the network status word. No conditioning of the signal is provided.

Most sensor networks are designed to transmit bias power to the sensors so that their present status can be sensed without a separate source of power for each device. In most cases, there is a module at the network node that allows termination of more than one I/O point to share the cost of the node. Typically, this is convenient since sensors are frequently clustered together around a common piece of equipment.

Some sensor networks are wired in a daisy chain or multidrop topology to reduce the field wiring as much as possible. Other sensor networks are wired in a star topology to reduce the latency delays of sensing. Still other sensor networks are wired in a ring topology for network reliability. Figure 3-2 illustrates these network topologies.

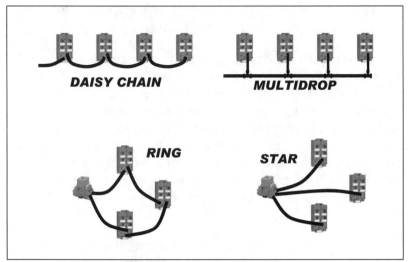

Figure 3-2. Network Topologies

3.1.1 Wireless Sensor Networks (WSN)

Much of the cost of installation of wired networks is for the wire itself. Installation of wire in a plant/factory environment is costly. The natural conclusion is to eliminate the wires by using WSN. The natural topology of a WSN is the mesh as illustrated in Figure 3-3. Notice that the sensors of a mesh also serve as communications hubs for devices that out of radio range to reach the gateway or host device. Also note that there can be alternative paths between any two devices. These are some of the advantages of WSN.

Installation of WSN also has costs not associated with wire and fiber optics. The ability to transmit and receive data over a radio (wireless) link is not always going to work with the same degree of certainty as a wired link. Atmospheric conditions such as rain, fog, or snow can dramatically affect transmission of wireless signals. Another problem can be traced to the "canyons of steel" that describe many process plants and even factories. When radio signals bounce off steel equipment, signals reaching remote devices must journey through a longer distance than the direct path. This is called a multipath signal and makes the signal taking the longer path arrive out of phase

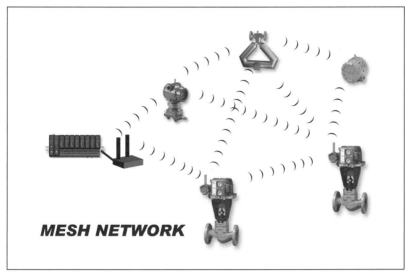

Figure 3-3. Mesh Network Topology

with the direct signal, resulting in a signal cancellation that is called "fade." Installation of WSN must account for multipath and the ability of signals to be received.

3.2 Fieldbus Networks

The original fieldbus was standardized by the ISA as ANSI/ISA50.02 beginning in 1992, with the final specification published in 1998. The process control version of this standard was implemented by the Fieldbus Foundation and is their H1 specification. On submission to the IEC for international standards, there was a great deal of controversy and debate, eventually leading to the adoption of seven more network architecture options in their standard IEC 61158, published in 2000. This IEC standard defines the term *fieldbus*. In this book, we have included in the scope of "fieldbus" all industrial networks that are designed for installation in the manufacturing plant or shop floor and in which there is distributed and programmable intelligence at each node of the network. Therefore, this classification includes all of those networks previously classified as *devicebuses*.

Often, it is necessary to measure the position of an object more precisely than present or absent. The most simple position measurement device is an LVDT (Linear Variable Differential Transformer) or where rotary motion is used, an RVDT (Rotary Variable Differential Transformer). These devices are trans-formers, which when biased with an AC current, measure motion accurately as a variable reluctance proportional to posi-tion. The reluctance is then converted to a digital value to be used by the controller. PLCs usually report such values as the output of the A/D (analog to digital) converter that occupies a register location. Other devices such as synchros and resolvers are also used to measure location or rotary motion in a similar manner. Optical encoders supply rotary position as a digital signal based on an encoding wheel that rotates with the shaft of the instrument.

Variables such as temperature, pressure, flow, level, current, voltage, and pH are measured by analog instruments. The term *analog* is still used for such measurements because they represent a scalar value, even when the underlying mechanism may be purely digital. Such measurements result in a digital value that eventually will be converted to *engineering units*. Devices that measure such variables are considerably more complex than simple digital discrete points, and often require many parameters to perform the scaling and filtering of the raw measurement. This leads to a requirement for bidirectional communications with these sensors. Foundation™ Fieldbus and Profibus-PA were designed for these "smart sensors." Additionally, many of the networks originally designed for communication of binary discrete sensor data can be adapted to transmit scalar values from scalar sensors and to scalar actuators.

Clearly, a different network is required to transmit digital dis-crete sensor data from that required to exchange parametric scalar data with smart analog sensors. It was for this task that fieldbus networks were created. The term *fieldbus* is used when a programmed device (microprocessor) is located at the net-work node, and the capacity exists to control its execution by downloading programs and/or configuration data. Although programs can be permanently stored in ROM, they can also be downloaded and stored for execution. The data for signal processing and perhaps control (called *configuration*) is also downloaded from a host computer. Data sets are then transferred to the host computer on demand, on schedule, or upon an exception condition. Communications with a smart sensor are truly a computer-to-computer data exchange across an information network that is called a *fieldbus*.

Smart sensor information is usually transferred in terms of a set of data sampled at the same time, which is called an *atomic data set*. Time coherency is important and cannot ever be accomplished by successive queries. Time synchronization is also important to dynamic control and cannot be achieved without a synchronous network. Dynamic control such as the PID (proportional, integral, derivative) algorithm used in process control, positioning control, robotics, airframe positioning, and motion control is based upon atomic data sets sampled at exact time intervals.

Many of the sensors or transmitters used in process control need electrical power for operation. This power previously was delivered on 4-20 mA analog connections, and now must be delivered using the fieldbus network. Often process control field instrumentation is installed in areas having the potential for combustible gases or dusts. These areas are classified as hazardous by the National Electric Code (NEC), NFPA[1]-70, Article 500. Even when field instrumentation is installed in explosion-proof or inert gas purged enclosures, electrical power carried on a fieldbus cable may present a hazard unless measures are taken to limit the electrical power using intrinsic safety devices called barriers. Intrinsic safety barriers installed on the fieldbus cable limit current flow to prevent the generation of a spark with sufficient energy to ignite a flammable gas, vapor, or dust mixture, specified by the area's hazard rating, if the cable is accidentally broken. Some data transmission methods are regarded as inherently safe, such as wireless, pneumat-

---

1.   National Fire Protection Association, Quincy, MA, USA, http://www.nfpa.org

ics, and fiber-optic sensors when energized by LED sources, not lasers. Because the fieldbuses intended for use in process control were to replace the 4-20 mA that delivered the signal and power on intrinsically safe 2-wire instrument cable, they also needed to operate with only two conductors on similar cable. Delivery of power to field sensors with intrinsic safety is one of the most difficult achievements of some fieldbuses.

While wired sensor networks deliver sensing power to the simple devices that they connect, fieldbuses designed for connection of discrete I/O need more controlled power for the microprocessors in the network nodes, as well as for sensing power. Many of these locations in the process industries will also require intrinsic safety specifications, while applications in discrete parts manufacturing, packaging, and assembly line control will generally not require intrinsic safety designs. Additionally, although a small number of sensors and actuators may be grouped together at one location, most fieldbuses intended for manufacturing and assembly line control must connect these devices along a machine or transfer line that can be quite long. The number of I/O sensors and actuators for discrete parts manufacturing, packaging, and material handling can often be very large, leading to the use of multiplexing nodes connecting many I/O at one network location.

Use of wireless connections for process data acquisition and control has already begun with non-standard and proprietary equipment. WirelessHART (Highway Addressable Remote Transducer) will soon be used widely in process applications, and ISA100.11a conforming networks will also begin to be widely deployed by 2009. Discrete parts manufacturing industry fieldbus applications are of great interest, but have not yet been developed as of 2008.

3.3   Control Networks

Control level networks are intended to allow control systems to connect with each other, to serve as the path for connection of fieldbuses to control systems, and for control systems to

connect to business systems. Because large amounts of data
may be passed through these networks and message lengths
tend to be longer, data transmission rates tend to be faster than
with fieldbus networks. However, since they can be used to
pass time-critical data between controllers, control networks
must also be deterministic and meet the time-dependent
(usually called "real-time") needs of their intended
applications. Determinism in a network context is defined as:
there is a specified *worst-case delay* between the sensing of a
data item and its delivery to the controlling device. Real-time
in this context is defined as "sufficiently rapid to achieve the
objectives of the application." These are separate but
complementary requirements.

If the same control network is used to exchange both real-time
data between controllers and business information between
controllers and business systems, clearly there must be some
way to prevent business information from interfering with
real-time deterministic response. Many complex protocols
have been constructed for this purpose, but most control networks rely only on the underlying nature of the chosen network protocol. Usually determinism is achieved by preventing
message collisions, limiting the maximum message length, and
using high speed.

The benefit of using a standard low-level network protocol
such as TCP/IP over an Ethernet network is a lower cost. By
simply selecting standard Ethernet cabling and using full-
duplex Ethernet switches instead of passive hubs, a control network built on this commodity technology can guarantee that
there will be no network collisions, making such networks
deterministic. Using high speed such as 100 or 1000 Mbps and
the standard Ethernet maximum packet length of 1,500 bytes
means that other applications cannot "hog the wire," preventing time-critical data transfers. However, you still must do the
math! Remember that the definition of real-time and determinism requires that the network must make its bandwidth available for time-critical data transfers in less than the maximum
time period allowed for the control system. For example, if a
business application were to transfer a maximum size Ethernet

message (1,500 bytes) at 100 Mbps, the network would be blocked for a maximum time of about 150 μs. Normally this magnitude of delay is perfectly acceptable for both process control and factory automation needs, but may not be acceptable for motion control or machine control.

It would be nice if control networks and fieldbus networks could not be used for the same applications, but they can. It would also be nice if control networks were always confined to a business or control room environment, but increasingly they are being extended to the field and shop floor. In some cases, control networks are being used in applications normally requiring a fieldbus. In fact, all of the control networks were developed from one or more of the fieldbus networks and use the same application layer and user layer protocols. Since control networks are related to fieldbuses, there will continue to be a very loose dividing line between them.

Wireless control level networks are already in common use, when the popular IEEE 802.11a/b/g/n standard, also known as Wi-Fi, is also referred to as "wireless Ethernet," is used. Since most of the popular control level networks are based on Ethernet, substitution of Wi-Fi for any or all segments is easy and does not require any change in the application and user layers. Often substitution of Wi-Fi for any control network segment is made by the user, and will usually work well, but may not be supported by any of the standards for control level networks.

3.4   Safety Buses

When safety interlocks, emergency-stop (E-Stop) buttons, and high-level alarms were hardwired to pumps, motors, and other devices, we were sure that they would work all of the time. However, when these switching devices are connected via a sensor network or a fieldbus, we are no longer guaranteed that a change in switch status could activate the safety condition *if there was a network interruption or delay*. This creates a new requirement—safety buses must not only be deterministic, fast enough, and reliable enough, but they must also provide some affirmative action indicating a failure of the safety bus node to communicate. Only then can we use safety devices on sensor networks and fieldbuses.

Any sensor network or fieldbus can be used for safety detection, interlocking, and shutdown if it has a method to *quickly* detect a failure of the network or of the node to which the safety sensor or actuator is connected. *Quickly*, in this context, is defined as fast enough for the safety function. Although hardwired safety interlocks can be thought of as instantaneous, some reasonable time delay between activation of the interlocking signal and the safety actuation device can be specified for a safety bus. Any failure of the network or the safety nodes should then cause an immediate actuation of the safety mechanism to create a *fail-safe* condition. The *fail-safe* mechanism must be at the actuation device, not a command sent via the network.

The Safety Bus function is often confused with high-reliability networks that are achieved with redundant wiring paths and often redundant network nodes. The word *redundant* in the context of networking means that there are two or more routes between any message sender and the intended recipient. There are many ways to achieve network redundancy including dual redundancy, bidirectional ring topology, resilient ring structures, and grid or mesh network architectures. All of these are to be found in industrial automation networks, but *high reliability alone does not make a network suitable for safety*. Likewise, safety buses do not need to be highly reliable; they need only to cause the desired trip function rapidly if they fail. When redundant networks are used for safety, they simply make network failure an unlikely source of a safety trip.

Review Questions, Industrial Network Basics

1. What are the layers above the ISO/OSI seven-layer communications stack?

2. What is the primary purpose of a sensor network?

3. What are the four topologies used for networks?

4. What makes a fieldbus network different from a sensor network?

5. What is the primary difference between a fieldbus and a control network?

6. What is the primary action of a safety bus?

3.5   User Layer Protocols

The top of the ISO/OSI seven-layer communications stack illustrated in Figure 3-1 is called the application layer, which serves to make all of the network services available to the applications programmer. Additional layers above the application layer are often called *user layers* that add functionality by making network services *most often needed by a class of user* more easily available. All popular fieldbuses and control networks offer user layers for their intended users, but sensor networks are so simple that user layers are not always provided or required. User layers are termed this because they were not named by the ISO in its design of the seven-layer OSI stack. In fact, the user layer may itself be composed of several layers. A typical industrial automation network will use two different user layers for its own applications: one for the efficient cyclic data acquisition and control data transfers and another for the information data transfers needed for operator display and interaction. There are user layers constructed for some vertical industrial applications as well.

3.5.1 Real-Time Data Acquisition and Control

Real-time data acquisition for control purposes will use a deterministic network architecture such as Modbus, Foundation Fieldbus, Profibus, CIP (common to DeviceNet, ControlNet, and EtherNet/IP), or iDA. In most cases, these networks use the services of a corresponding network application layer to access data from the underlying network. Use of these real-time layers usually requires that the software used for the application itself is highly aware of the user layer upon which it is depending.

Sensors and actuators used for discrete control terminate in a multiplexer of a PLC or a remote I/O unit. The term *multiplexer* is used generically to mean an electronic device to sense the status or position of an input sensor or to change the state of an actuator. The status of inputs and the state of outputs appear in a multiplexer as a register or a word of 16 binary bits where each bit represents an input or output (I/O) point. Process control likewise depends upon data from sensors being multiplexed and sent by the network to process controllers and the resulting output values being sent to actuators. The role of the real-time network is to transfer the status words to the PLC or sensor data to the process controller rapidly enough for it to complete its scan and control cycle and then to transfer the output status words or values back to the remote I/O registers or process actuators, all within a specified time period.

All of the sensor networks can meet the requirements of real-time data acquisition and control that are consistent with the timing of their networks and the number of I/O points that they are expected to scan. Very little can interfere with the operation of a sensor network.

Fieldbuses can also meet the real-time requirements for data acquisition and control as long as the definition of real-time is consistent with the timing of the control cycle and the network timing, as well as the number of I/O points to be scanned. However, fieldbuses must be configured so that the PLC or process controller can regulate network access to ensure that all scan and control cycles can be completed within the specified time period. This is the requirement for deterministic data transfer that cannot be met by uncontrolled networks such as those used for information access.

All of the control networks can also meet the real-time requirements for data acquisition and control, but usually the timing is much more relaxed than that of a fieldbus or a sensor network. The difference is usually in the protocol of a fieldbus that has its time synchronization at layer 2 (data link) of the network, as opposed to control networks that achieve timing at layers 3 and 4 (network and transport) of the network.

3.5.2 OPC and Information Access

Many applications in industrial automation are highly generic and intended to operate with a variety of automation systems. Originally, authors of this type of application software were required to create *drivers* for each of the automation systems to which they would connect. OPC (object linking and embedding for process control) fixed that situation by establishing a common interface (OPC *client*) for any application working with automation systems supplying an OPC *server*. Communications would then occur only between the OPC client and server, freeing the software supplier from the particularities of the automation system.

OPC was originally designed for operation on the Microsoft Windows® operating system. It is founded on the Microsoft Component Object Model (COM) and its network distributed (DCOM) equivalent. Using COM, two applications running under Windows on the same PC can pass data to each other as messages between objects. The same applications may also pass messages to each other when they are run on different PCs on the same network using DCOM. COM is based on a common operating system feature called RPC (Remote Procedure Calls) that is implemented in Microsoft Windows as its message-passing interface between executing objects and is called OLE (Object Linking and Embedding.) When the two objects are executing on different computers, the Windows operating system uses the user-defined network protocol to communicate. Microsoft supports TCP/IP as the standard network protocol, but any protocol defined in the Windows Network Connections may be used. If the networked device is not running Windows, it must then support COM/DCOM, or SOAP (originally an acronym for Simple Object Access Protocol, but now does not mean anything) that is now an open standard supported by the W3C (Worldwide Web Consortium), a standards body responsible for Internet standards. Microsoft and many others support SOAP on many other operating systems, including several used for embedded systems appropriate for industrial automation devices.

The initial OPC protocol was called DA (Data Access), which used very simple data structures. The data understood by DA is a simple register, 16 bits long. The meaning of the bits is to be defined by the end user during the application. Although DA was sufficient for typical PLC applications, process control functions required additional data structure for control loop and instrumentation attributes or parameters. At first, OPC created XML (eXtensible Markup Language) schemas for defining these data structures as a layer to be implemented on top of OPC/DA. Before OPC/XML could be widely used, it was recognized that a stronger object attribute definition method was needed. This was introduced as OPC/DX (Data eXchange), again built as an additional layer operating above OPC/XML. The purpose of OPC/DX is to allow the definition of data independent of both the control system supplying the server and the data management or presentation system supplying the OPC/DX client.

OPC has now evolved to OPC/UA, or Unified Architecture. For a complete description of OPC/UA, the reader is encouraged to read the information that can be obtained in a web search for OPC/UA. These changes in OPC were made necessary by the facts that support on platforms other than Windows has become necessary, and that the original COM, DCOM, and OLE roots have become obsolete. OPC/UA is based upon a series of open, IEC, and Internet standards that support distributed object-oriented data transfer such as SOAP and EDDL (electronic data definition language.) While the mechanisms of OPC/UA are different from earlier versions, the objectives are the same: to provide a common interface between data sources and the users of that data.

3.5.3 Microsoft .NET Architecture and Java

Microsoft has replaced its entire object modeling system and many other aspects of its architecture with .NET (pronounced dot-net) Framework. The intent of .NET is to establish a single consistent program execution framework independent of programming language. Microsoft calls this the CLR (Common Language Runtime) environment, which is a virtual machine. Microsoft has standardized the .NET Framework specifications through ECMA (European Computer Manufacturer's Association), an open standards body. This concept is almost the same as Sun's JVM (Java Virtual Machine), but done five years later and with many Windows advantages. Sun's JVM interprets original program language statements compiled into *byte code*, which are the instructions to the JVM. .NET is slightly different—original programming language statements must be written to the .NET class library standards, and then are compiled into the target machine code, with most of the work done by the class library. Microsoft claims that .NET will run faster on all targeted computers, but Sun counters that Java runs on any computer supporting the JVM. Microsoft has support for the JVM machine in their own version of Java that runs under all versions of Windows. In fact, there will be almost no observable differences between .NET and Java. The biggest difference is that .NET supports a smooth migration path from COM/DCOM to the .NET object communications. This means that .NET supports OPC. Microsoft's version of Java also supports OPC/UA.

The .NET object access method is called SOAP, which supercedes COM. SOAP is an XML-based standard of the W3C (World Wide Web Consortium), one of the Internet standards bodies. The effect on OPC is to free it from the Microsoft operating system constraint. The OPC/UA specification was based on the use of .NET/SOAP.

3.5.4 FDT (Field Device Tools)

Most of the fieldbuses define attributes of devices used for sensing and control. These attribute definitions are contained in user layer specifications. Table 3-1 lists the name of the specifications for the fieldbuses and some of their properties.

FDT was created to eliminate the need for the user to maintain the different attribute definitions for HART, Profibus-PA, and Foundation Fieldbus. FDT allows the field device supplier to offer a single DTM (Device Type Manager) independent of the fieldbus to be used for a project, whereas the host device uses

Table 3-1. Fieldbus Attribute Definitions

Fieldbus	Attribute Definitions	Typical Use
Foundation Fieldbus, Profibus-PA and HART	EDD (Electronic Data Definition)	Process control loop and instrumentation attributes.
DeviceNet, ControlNet, and EtherNet/IP	EDS (Electronic Data Sheet)	Defines the attributes of the device

an FDT Framework server. Most process control suppliers seem to be supporting FDT, especially those that also support Profibus-PA. FDT does not have this same level of support in the factory automation world since few PLC suppliers have created FDT Framework servers or DTMs for binary field devices. It seems that FDT and OPC/DX are two very similar methods of solving the same problems.

OPC states that it is a high-level protocol for standardizing host to controller communications. FDT claims to be a high-level protocol for standardizing field device to controller communications. Both are built on Microsoft COM/DCOM and use XML-encoded data frames (in OPC/DX, XML, and UA). OPC states that it is intended for Ethernet-based automation networks, whereas FDT is designed for fieldbus networks. Clearly, as Ethernet-based control networks converge into the domain of fieldbus, there will eventually be a conflict between OPC and FDT/DTM.

Another potential conflict for FDT is the standardization of EDDL by IEC 61804. The developers of FDT state that it was specifically developed to resolve the engineering problem of defining device attributes for smart devices. This is one of the goals of EDDL as well as to provide real-time high-speed data access to device data, which was not a goal for FDT.

The major advantage of FDT over both EDDL and OPC has been to give the field device supplier the ability to construct comprehensive visualization tools for detailed analysis of the data contained in the field device, and especially for use in calibration and diagnostics. These tools, using FDT, are independent of the control system supplier.

In 2007, the FDT Group joined the EDDL Cooperation Team (ECT) with the purpose of working out the differences and avoiding end user confusion over these two approaches. The EDDL work of IEC 68104 has since become ANSI/ISA104 while the FDT specifications are still on a path to become ANSI/ISA103. Many of the graphic capabilities pioneered by FDT for display of local instrument data have now been added to EDDL.

3.6   Convergence and Downward Migration

The end user should not care about the underlying network protocol for an automation network, but only about the difficulty and cost of installing the automation system and the support of the network by the suppliers. Most of the work of EDDL, FDT, and OPC has been to build a layer of abstraction above the network protocol to make it transparent to the end user and the system suppliers. However, the end user will still be concerned with installation. Many of the more modern network protocols are directed toward reducing the cost of installation by using commodity network components and wiring.

Eventually, proprietary networks will be fully displaced by industrial networks conforming to international standards, or at least industry-accepted "open" specifications. Most of the differences between networks originally designed for factory automation and those created for process control will begin to disappear except for the highest levels of the user layer. For example, there is practically no difference between the installed cost of Foundation Fieldbus HSE, EtherNet/IP, or PROFInet, each of which uses standard commercial off-the-shelf Ethernet wiring and components. However, the user layers are quite different.

3.6.1 Wireless on the Shop Floor

The reduced cost of Ethernet-based networks is driving this fast, low-level, and low-cost technology to the field or shop floor. Another Ethernet side effect can be seen in the application of wireless technology in the Wi-Fi group of wireless protocols. Wi-Fi is essentially wireless Ethernet. Any higher level application layer and user layer can communicate via Wi-Fi at data rates up to 600 Mbps, without knowledge of the fact that it is on a radio link. Wi-Fi is the most common wireless technology in 2008; however, it has significant problems for operation in the electrically noisy environment of a process plant or the shop floor in a manufacturing factory.

The popular Wi-Fi-a/b/g standards can achieve a theoretical maximum of 54 Mbps using one of the 2.4 GHz channels. However, the not yet finalized Wi-Fi-n standard, which has a maximum rated specification of 600 MHz, is even more interesting. Wi-Fi-n allows the bonding of several radio channels, including the channels in the 5 GHz, band to achieve its high speed, but is not due to be finalized until the end of 2009. The feature of Wi-Fi-n that is most appealing to industrial use is its adoption of MIMO (multiple input, multiple output) technology. MIMO has the demonstrated potential to eliminate the adverse effects of reflections that cause multipath distortion appearing as signal fade. MIMO achieves improved reception through detecting the multipath signals and either eliminating them or phase-shifting them to amplify the received signal. Early experiments suggest that using Wi-Fi-n can achieve excellent behavior in both process plants and factories notorious for their "canyons of steel," the cause of poor performance of Wi-Fi-a/b/g. Both commercial and industrial versions of "pre-N" devices are already widely being sold. Of course, use of wireless technology brings problems of security and privacy to industrial networks, which was never much of an issue before.

Wireless technology provides excellent solutions to the problem of the high cost of industrial wiring, and also provides an
ultimate barrier to electrical surges introduced to field equipment through field wiring. The cost of these wireless advantages is the difficulty of supplying power to field devices,
which previously were powered by the same cable used to conduct the data exchange between the field device and a host system. Although there are a few wireless field sensors offered for
sale in 2008, it is likely that the potential cost reduction of
avoiding wire/cable installation and maintenance will provide
an expanding market for wireless sensors in the future. Additionally, new applications for industrial measurements and
controls are being found for wireless devices. These applications were often not economic when they required wired connection.

As of the end of 2008, a few WirelessHART process field transmitters are being offered commercially. In addition, a simple
device to convert wired HART transmitters and valve positioners to the WirelessHART protocol is available. WirelessHART is
a defacto standard supported by the HART Communications
Foundation, and is an IEC PAS (publicly available specification.) An IEC working group to develop a international standard for WirelessHART has been approved.

WirelessHART uses the IEEE 802.15.4:2006 standard modified
to hop among the 15 or 16 frequencies (channels) specified by
that standard in the 2.4 GHz ISM (industrial, scientific, and
medical) band. The slot time is constant for the entire network,
usually 10 ms. Transmission is encrypted using a 128-bit key to
achieve a high degree of security. Field devices are all part of a
mesh network (see Figure 3-3) with a secure method of building and repairing the mesh. The advantages of a mesh network
are redundancy, increased total distance, and removal of the
line-of-sight restriction.

A simple transport layer is defined to ensure end-to-end message delivery and confirmation when required. Like wired
HART, data access is polled using HART commands, including
all maintenance functions of the WirelessHART network. WirelessHART devices may also be set to transmit data using a pub-

lishing method. WirelessHART devices are provisioned (network setup) through a wired connection.

WirelessHART was developed to provide a simple wireless network for field instrumentation, and to enable a wireless method to access diagnostic data in HART instruments installed in the past. It has been estimated that there are more than 25 million of these HART instruments that can only provide their digital diagnostic data to a hand-held terminal, since the control systems to which they are connected cannot access that data over the connecting 4-20mA wire. WirelessHART is seen as an answer to this problem.

ISA standards committee 100 has been developing a comprehensive standard for industrial wireless communications. The first release from this organization is ISA100.11a, intended for process data acquisition and limited control needs in the process industries. While the ISA100.11a standard has achieved ISA and ANSI standard status, it has not yet passed international (IEC) standardization. The fact that WirelessHART and ISA100.11a are addressed to the same market has been noted by many users and the IEC as well. They are technically similar, but very different in detail, causing many users to request that they be merged. ISA100.12 has already been formed to achieve convergence between WirelessHART and the ISA100.11a specification. At the time of this writing (2008), the standardization path for wireless process control and factory automation networks has not yet been finalized. Resolution between WirelessHART and ISA100.11a may be completed in 2009, but standardization is still unresolved.

ISA100.11a also uses the same IEEE 802.15.4:2006 standard in a way similar to that of WirelessHART by hopping among the 15 or 16 channels in the 2.4 GHz ISM band. ISA100.11a has far more options to adapt its network for a wide variety of applications including segmentation of the network and peer-to-peer messaging. Unlike WirelessHART, each network segment may use a different hopping pattern and its own allocated time slot. These choices were made to allow large networks to be formed where segments may overlap. Field routers have also been defined to reduce the number of hops required to reach the host device, and to bridge geographically separated network segments.

ISA100.11a also uses mesh networking, but allows devices at the edge of the network to not route messages for other devices. This can increase security by preventing unauthorized devices to access plant networks from outside the plant. It can also reduce the cost of devices by making them simpler. Additionally, ISA100.11a uses Internet-conforming IP addressing to make data from field devices addressable remotely. The Transport Layer implements secure end-to-end message delivery and confirmation and is based on the use of Internet-conforming UDP (User Datagram Protocol) messages.

The Application Layer of ISA100.11a is completely object-oriented, in which data in field devices can be addressed using IEC 61804 standard EDDL protocol. For networks not using this standard, all messages may be encapsulated and tunneled to the requesting host device.

ISA100 has already agreed to extend its work in several ways:

- New physical layers (chip technology)
- A wireless or wired backbone (backhaul) network
- Factory automation sensor network applications
- High security or trustworthy applications
- RFID and Location-based services

ZigBee is an organization specifying additional higher layer protocols to use the same wireless IEEE standard, numbered 802.15.4. It was designed to operate on the shop floor and to avoid interference with Wi-Fi. It is also low cost, requires little power, and can transport Ethernet messages. Although ZigBee may operate with a star topology like Wi-Fi, it also allows operation in a mesh network topology as previously illustrated in Figure 3-3 (see page 15).

By the next revision of this book, we should have a much clearer picture of future industrial automation networks. Wireless offers too many advantages to ignore and will tend to change all of the networks being reviewed in this book to "legacy" status. However, you cannot install standard wireless networks today, but you may purchase WirelessHART devices and other devices that claim to be upgradable to the ISA100.11a standard. Wireless networks of any kind do not yet exist for factory automation sensors. Therefore, the remainder of this book will concentrate on real networks with real devices. I hope you enjoyed a glimpse into the not-too-distant future.

Review Questions, Network Architecture

1. What are the uses for a user layer protocol?

2. Explain determinism. Why is it important for data acquisition and control?

3. What is the purpose of OPC/DX?

4. Why is Microsoft .NET architecture more important for control systems than Java JVM?

5. Field Device Tools are used for information access to what level network?

6. How are WirelessHART and ISA100.11a different?

Unit 4: Network Applications

In this section, there will be many applications mentioned for the purpose of illustrating the selection of a specific network architecture. In many cases, several different networks could be selected. In these cases, the benefits and disadvantages of each choice will be discussed. Often, there will be no obvious reason to select one over another except for support offered by the equipment supplier for the project.

Applications are divided into groups by the type of network that would be used. Every equipment supplier tries to extend the networks they support to all possible applications, but this will not be done here. However, for some applications, scaled-down protocols of other networks exist that enable them to be used for lesser-functionality applications. This analysis is generally free from commercial influence, but just because a particular network *can* be used does *not* mean it *should* be selected. Selection of a network architecture cannot be made in a vacuum. Support by your equipment supplier is very important, especially your supplier's local support. In many cases, the technically superior network is *not* the best choice if it is not well supported.

Network support means that both ends of the network can be used. When networks were just wires and electrical signals, support was relatively easy. Now that networks have much more complicated higher level protocols and even user layers, software support and configuration support for devices become critical.

4.1   Sensor Network Applications

Sensor networks are designed for the simplest applications.
How do I connect limit switches, solenoid valves, pushbuttons,
indicator lamps, and alarm displays? Sensor networks are
designed to reduce installation cost by reduction or elimination
of direct wiring to connect these simple devices. There are four
situations that must be treated differently:

- Devices are clustered at one location

- Devices are clustered at several locations around the
  process or machine

- Devices are distributed along the length of a process or
  machine

- Devices are clustered at several locations along the
  length of a process or machine.

One of the problems of networking sensors is the cost of the
network interface – the electronics necessary to convert the sensor state to a corresponding bit on the network or to convert a
bit on the network to the desired state of an actuator device.
Although it is possible to build this interface into the device, it
has generally been more economical to share the network interface with other similar devices. If the devices are clustered
together at one location or clustered at several different locations, then sharing makes economic sense. If the devices are not
clustered but distributed individually along a machine or a
process, then using devices with embedded network interfaces,
or perhaps not using a sensor network at all, may make economic sense.

Wiring simple switching devices to either a sensor network or
to the remote I/O of a control system still requires copper
wires. The cost of copper wiring does not end with the cost of
the wire. The wire must be pulled or installed between the
device and the network interface or the remote I/O unit. The
wire is usually installed in a conduit or raceway, apart from
power wiring to avoid induced currents and noise. Each end of
the wire must be connected to the designated device and to the

assigned network interface point or remote I/O termination. There is also the cost of designing and drawing the connection diagrams and wire lists. Finally, each point must be tested or "rung-out" manually. Every process change that involves changes in the devices must also be documented as part of ongoing plant maintenance. When the wiring between the device and the network interface is very short, there is a mea-surable saving in wiring, installation, testing, and ongoing maintenance. This is the cost benefit of clustering and using a sensor network.

This is where the question usually occurs: Can we use a wire-less network? The incentives are strong to eliminate the wiring, installation, and maintenance. Testing would still be required to assure proper reception of the wireless signal. To date, there are no economic wireless solutions for discrete devices, but look for wireless connections to device cluster interfaces in the near future.

Which sensor network should I pick? Or, should I use direct wiring without a sensor network? To make such a decision, one factor is installed cost. You would need to estimate the installed cost of each method, including the cost of the I/O cards/mod-ules and the network scanners or interface modules. Most of the time, the fact that one sensor network can connect 32 or more I/O points will make it the lowest cost method by reduc-ing the number of interface cards in the I/O rack. The cost of multidrop or daisy-chain wiring is usually lower than the star wiring required for direct wiring installation. For most new installations, a sensor network will be used without making a detailed cost justification. The savings in I/O interface cards alone are usually enough to make the choice self-evident. How-ever, when I/O speed must be extremely fast, as for machine control applications, direct wiring to remote I/O is usually jus-tified.

All sensor networks operate similarly by making the I/O appear in a series of registers in the I/O unit in much the same way as a directly wired I/O appears as a series of registers of the PLC or its remote I/O unit. This mechanism makes the presence of the sensor network invisible to the PLC, requiring no software to operate. When the control logic is implemented, the registers of the sensor network are treated as though they are the registers of a conventional I/O card with many more points. For example, a conventional I/O card may connect 8, 16, 32, or 64 I/O points. When this card is read, the I/O points will appear to be mapped to bits of a single register of 16-bits, 2 registers, or 4 registers as required. Similarly, when a sensor network connects the same number of I/O points, it is mapped in the scanner for the network into 1, 2, or 4 registers. Using this scheme, typical PLC programming will not need to distinguish differences between conventional I/O and sensor network I/O, as far as addressing is concerned.

One difference between sensor network I/O and conventional I/O remains. Conventional I/O cards are usually dedicated to a single function, such as input or output. Sensor networks usually allow each I/O point to be either input or output or allow some dedicated mix of inputs and outputs at each network node. This will not make any difference to the user as they program the logic, as long as they note which points are for inputs and which are for outputs. Most sensor networks and conventional I/O allow the actual state of outputs to be read as though they are inputs. This makes the mapping of I/O points different between direct-wired I/O and sensor network I/O as illustrated in Figures 4-1 and 4-2.

In Figure 4-1, addressing of the I/O points depends first on the register to which the I/O card is connected and then to the point number on that I/O card. Addressing in Figure 4-2 depends upon the I/O point of the sensor network termination to which the device is connected and the number of I/O supported by the sensor network device. There are no particular advantages to either, but they are quite different.

Sensor networks were originally created for factory automation applications in which most of the sensors and actuators are binary 2-state devices, such as limit switches and solenoid valves. In many of these applications, speed of detection is critical, meaning a change of state must be detected within 1 scan cycle of a PLC. To most designers, this means approximately 3

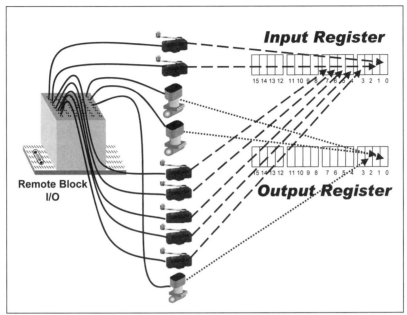

Figure 4-1. Direct Wiring of I/O and Register Mapping

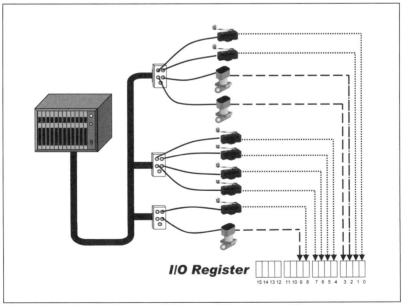

Figure 4-2. Sensor Network Wiring and Mapping of IO

to 5 ms for a typical machine control application. Sensor networks were designed to meet these speed requirements.

The number of discrete I/O points required for a particular control element (rung of ladder logic) is typically small, but the number of interrelated control elements for a given manufacturing process is usually very large. A typical machine control or materials handling application will have hundreds or thousands of discrete I/O points and typically few, if any, scalar (analog) measurements. Discrete sensor networks are focused on these applications.

4.1.1 Factory Automation Sensor Network Applications

In the automotive industry, many operations are performed on subassemblies that are supplied to the overall assembly line for automobiles. In the automation of subassembly manufacturing, the PLC is the primary controller, and discrete sensors and actuators are the means by which human labor is replaced. The process described here is very typical of this industry and is the process of assembling a brake drum, consisting of a finished casting into which 4 to 6 lug bolts are inserted. Usually, these brake drums are manufactured to order by several machines in parallel, each supplied with a fixed set of raw materials and producing only one product. Figure 4-3 is a photograph of a brake drum assembly machine, showing the connection of the AS-interface sensor network to the limit switches, electric motors, and solenoid valves that power the pneumatic actuators.

The PLC is mounted in an equipment rack some distance away and is responsible for controlling this assembly. It activates the mechanism to move a new brake drum casting into the proper position, insert each of the lug bolts, and tighten them to the preset torque. Next it inserts the correct wheel bearing assembly and fastens it in place. Finally it moves the assembled brake drum to the conveyor belt for inspection, packaging, and shipping. Each brake drum assembly station is set up for the insertion of up to 6 lug bolts into a variety of casting diameters.

Figure 4-3. Brake Drum Assembly

These manufacturing stations are designed for 8 limit switch inputs, 2 solenoid valve outputs, and 2 electric motor starter outputs. The 12 I/O points are connected through AS-interface modules, which are interconnected to the PLC with AS-inter-face flat cable, which also supplies sense power for the inputs. This illustration shows some of the rugged wiring typically used for both sensor network and fieldbus. The wiring from each sensor, solenoid valve, and motor starter terminates, in this example, in an M12[1] round 4-pin plug, which is then con-nected to two sensor network interface modules each contain-ing six M12 receptacles for inputs and outputs.

Manufacturing of discrete parts, typical of the automotive industry, uses processes such as stamping, casting, machining, painting, and testing. Electronics manufacturing involves inserting components onto a printed circuit board, soldering, cleaning, and testing of the finished board. White goods manu-

1. IEC 61076-2 (1998-12) Connectors for use in d.c., low-frequency analogue and digital high-speed data applications - Part 2: Cir-cular connectors with assessed quality - Sectional specification

facturing usually involves stamping, painting, and testing. All of these industries must assemble the components they build into their finished product. The assembly line is where products are put together one step at a time. Materials handling is the primary process of the assembly line and involves conveyors, deflectors, robots, manual workstations, and automated attachment and testing. Most of the operations involve sensing for the presence of a component, validation of its identity, and activation of the manufacturing operation. Even for processes that are mostly manual, there is usually a materials handling operation that must be synchronized using logic.

PLCs typically perform the logic of materials handling, depend upon sensors to supply the data, and use actuators to carry out the operations.

The materials handling station illustrated in Figure 4-4 shows a conveyor belt used to move the work-in-progress to a number of assembly stations. There is a photoelectric sensor to determine when an object is on the conveyor and a laser barcode reader to determine if the object is to be assembled at this workstation. If these conditions are true, then the diverter is activated to move the object to the workstation for assembly. Otherwise the object continues on the conveyor belt to be serviced by another workstation. The photoelectric sensor typically needs 1 bit of digital input, the diverter needs 1 bit of digital output, but the barcode reader can read alphanumeric

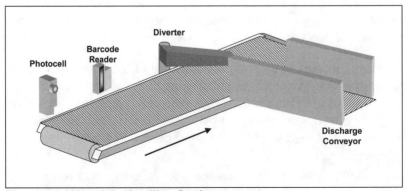

Figure 4-4. Materials Handling Station

barcodes up to 16 characters or 128 bits of data. Since sensor networks are designed for single-bit devices, and some 2- to 4-bit devices, use of a fieldbus would be a better choice. In this case, a hybrid approach in which there is a fieldbus network node for the barcode reader with a sensor network interface for the discrete I/O would be a good choice. Both Profibus and Interbus offer this configuration with AS-interface sensor network modules fully integrated at the local node. In this case, however, with only two discrete I/O points at the node, it may be less expensive to directly wire them to a remote I/O block at this location.

Now the choice of sensor networks must be made. Details of each of the sensor networks are found in Chapter 5. In this chapter, we will look at some of the factors that will help make the decision for a variety of applications.

4.1.2 Process Control Sensor Network Applications

Only recently has the process control community focused on the use of sensor networks to achieve the goals of installation cost reduction for discrete I/O. In most cases, the speed requirements for discrete I/O have been much less demanding than for machine control; however, it was viewed that using an existing sensor network technology made more sense than adapting an existing process control–oriented fieldbus or developing a new sensor network only for process control. This means that sensor networks are important methods of cost containment even for process control.

The first process control applications for which sensor networks were desired were batch processes that typically have high discrete I/O counts, as well as modest scalar I/O requirements. Only recently has the focus expanded to continuous process control as automated startup, emergency hold, and shutdown applications are beginning to emerge. Previously, these operations were manual and required operators to go to the field to shut off or open blocking valves and startup pumps and other motors. Safety considerations are now requiring that these manual operations be validated by the control system, as

well as requiring automated operations to reduce elapsed time
from the decision to the execution. With automated startup,
emergency hold, and shutdown, all processes become batch.

In many continuous processes with control valves for each control loop, there will usually be full-open and full-closed limit
switches on the valve position, and switches on the blocking
and bypass valves as well. This translates to between 5 and 8
limit switch I/O for each control valve. A typical chemical or
petroleum refining process unit may have as many as 100 control loops, which means between 500 and 800 discrete I/O.
Batch processes or continuous processes that have automated
startup and shutdown will often have many more discrete I/O.
Figure 4-5 illustrates a process control application.

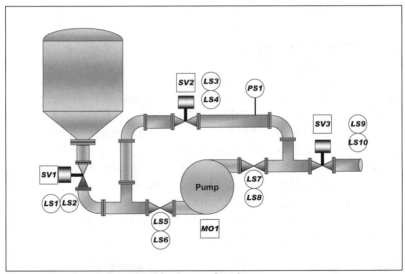

Figure 4-5. Batch Reactor Discharge Section

The illustrated portion of a typical batch process shows the
reactor discharge valve, the discharge pump, and the pump
bypass/recirculation line. Here, for simplicity, the automatic or
remote control valves are shown to be solenoid valves,
although they might typically be ball valves operated by a
pneumatic ball valve actuator that is controlled by a solenoid
valve in the air-supply line. These discrete outputs are shown

symbolically as squares with the XV label. There are also manual blocking valves to isolate the pump during maintenance. Each valve, including the blocking valves, is equipped with two limit switches to detect the full-open and full-closed position of each valve, as indicated by the circular elements with the ZS (Limit Switch) designation. There is also a pressure switch, the circular element labeled PS1, to indicate that the discharge pump is running at some preset pressure. There is a defined startup sequence to discharge the batch reactor (simplified in the following example):

1. The positions of the blocking valves are tested by checking ZS5 to ZS8.

2. If the blocking valves are fully opened, the output valve XV3 must be closed and the bypass valve XV2 must be opened.

3. Once XV2 and XV3 have been confirmed to be in their desired status by checking ZS3, ZS4, ZS9, and ZS10, then the pump drive motor MO1 can be turned on and the reactor discharge valve XV1 opened (prevents damaging pump).

4. After confirming that the discharge valve is opened by checking ZS1 and ZS2, confirm that the pump is running by testing PS1.

5. The output valve XV3 can now be opened, and the bypass valve XV2 closed.

6. When the pressure switch PS1 indicates low pressure, discharge has been completed and the discharge shutdown sequence can be started.

4.1.3 Picking the Sensor Network

Since there are several choices of sensor networks, what are the criteria that should be used to "pick the right one?" First, there is no "right one." There is only the one that is right for *your* application. Complicating this even more is that several of the

fieldbuses have versions that have been scaled down so that
they can be used as sensor networks. In general, if there are
needs for both sensor networks and fieldbuses, then you
should consider a sensor network compatible with the fieldbus.

The following sections look at each of the sensor networks and
the scaled-down fieldbuses from an application viewpoint.
This is not a study of protocols, but more a study of the wiring
requirements for installation of each bus.

Which Sensor Networks Does the I/O Manufacturer Support?

Although you may look at this question in reverse and pick the
sensor network first, that is most unlikely. Most of the time,
you will pick the I/O system as part of your PLC or other control system for reasons of programming flexibility, compatibility, features, cost, delivery, and overall vendor support. Then
you will need to determine which, if any, sensor networks the
suppliers support with scanner cards or modules. It is also
common for the sensor network supplier to merchandise scanners for a variety of PLCs. Often, the web site supporting each
sensor network provides a listing or a link to manufacturers
supplying compatible, and often tested, devices.

There is a large overlap between sensor networks and the low
end of the fieldbus spectrum. Many control system manufacturers do not support any of the true sensor networks, AS-
interface and Seriplex, but choose to support simplified versions of some of the fieldbuses such as DeviceNet, WorldFIP-
I/O, SDS (Smart Distributed System), and other versions of
CAN. Additionally, LonWorks can qualify as either a sensor
network or a fieldbus, depending upon its configuration, but
none of the major control systems suppliers support LonWorks
for their industrial automation products. The following sections sketch how a sensor network or a scaled down fieldbus
can be used to solve the application problems previously
described.

AS-i (AS-interface)

AS-interface is the most popular sensor network for connection of discrete I/O to controllers, to remote I/O, and even to some fieldbuses. AS-interface offers a low-cost multidrop bus topology, using either a 2-conductor round cable or a unique 2-conductor flat cable. Both cables deliver DC power to the node and the device using the same wires for both power and data. AS-interface taps are wired into the round cable and clamped onto the flat cable with vampire taps penetrating the insulation, with the tap forming a vapor-tight seal to protect the connection. Each tap provides an interface either to a device or to an I/O connection module that may connect up to 4 discrete inputs and 4 discrete outputs. The AS-interface bus is wired close to clusters of the I/O devices. A single AS-interface network can connect up to 124 inputs and 124 outputs and can be about 100 meters in length. Many other options exist.

Figure 4-6 illustrates a common AS-interface module for 4 I/O and using round AS-interface cable. The M12 connection is gas-tight and oil and water resistant, and is classified as IP67[1] rated. The M12 electrical connection is also a pin-and-socket junction, forming a gas-tight electrical seal Sensors, motors, and solenoid valves are available with M12 receptacles allowing the use of prefabricated molded cables, eliminating costly cable termination and fabrication errors. M12 4-pin connectors are used in many sensor networks.

Figure 4-7 illustrates AS-interface terminations with M12 connectors but using the AS-interface flat cable. This termination block is fastened directly to the AS-interface flat cable using vampire taps that penetrate the insulation of the flat cable as illustrated in Figure 4-9. The mechanical form of the AS-interface flat cable fits only one way into a slot of the termination block to prevent polarity errors.

Figure 4-8 illustrates a common AS-interface module intended to be used with conventional wired sensors and actuators. Wir-

---

1.   IEC 60529 (2001-02) Degrees of protection provided by enclosures (IP Code)

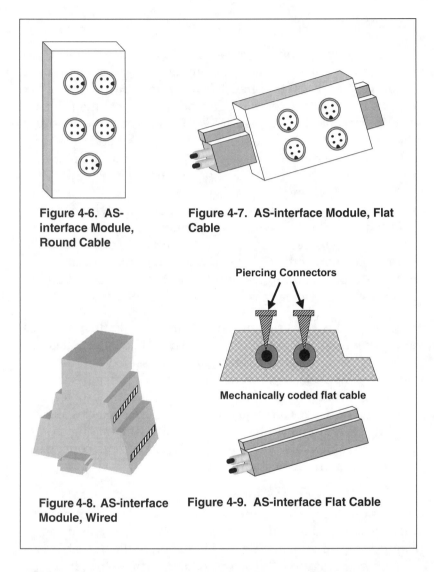

Figure 4-6. AS-interface Module, Round Cable

Figure 4-7. AS-interface Module, Flat Cable

Piercing Connectors

Mechanically coded flat cable

Figure 4-8. AS-interface Module, Wired

Figure 4-9. AS-interface Flat Cable

ing connections are made to bare wire inserted into the termination block and sealed with a screw compression mechanism. AS-interface round cable is likewise connected using a compression screw terminal.

Installation of AS-interface is very simple, especially when the flat cable is used, and sensors and actuators are purchased with M12 connectors. The flat cable is routed close to the sensors and actuators on the production line. A 4- to 8-port interface module is clamped to the flat cable at some convenient location. Each sensor and actuator is connected to the AS-interface module using pre-made jumper cables with M12 connectors at each end. Care must be taken to wire inputs to the input connections and actuators to output connections on the AS-interface modules. Custom wiring to AS-interface bare-wire terminations may cost less for materials, but the labor and maintenance cost is much higher. Bare-wire AS-interface terminations should only be used for connections made inside a wiring cabinet and not exposed to factory floor or plant environments.

Finding suppliers of sensors and actuators equipped with M12 connectors meeting the AS-interface wiring specifications may limit the choice of devices. For example, Rockwell's Allen-Bradley brand and Cutler-Hammer's brand of limit switches, photocells, and proximity switches are all available with M12 connectors wired to DeviceNet specifications but not to AS-interface.

Seriplex

Seriplex is another sensor network very similar to AS-interface, but a single network can connect up to 496 inputs and 496 outputs with a maximum cable length of about 1500 meters. Seriplex is not supported by many control equipment suppliers, but is supplied by Square-D for most Modicon products of Schneider Electric as well as many Square-D and Telemechanique brand switches and motor controls. Although Seriplex was established as an open sensor bus, and has many installed points in service, the sponsoring organization has been terminated and the only suppliers are now from the Schneider Electric family of suppliers.

Seriplex is organized as a network of I/O multiplexing nodes connected with shielded cable. All connections are terminated in a 5-pin M12 connector or directly wired using an insulation displacement connector (IDC) for each module. M12 connectors are designed for IP67 harsh and outdoor environments, while the IDC method is designed for mounting within enclo-

sures. I/O modules are available for all types of digital discrete
I/O and analog I/O as well. While the Seriplex network cable
delivers bias power for discrete sensors, it is not designed to
deliver any power to outputs. The Seriplex bus is a 4-conductor
shielded cable with a separate drain wire. Seriplex does not
specify plug and socket prefabricated wiring for sensors and
actuators.

Schneider has pledged to support Seriplex indefinitely in
North America, but they supply and support AS-interface
everywhere in the world. Although there are advantages to
Seriplex, such as longer cable length and the ability to mix analog and discrete signals, it is probably more expensive to install
and maintain than AS-interface. In the future, it is likely that
Schneider will de-emphasize Seriplex in favor of AS-interface.
This makes AS-interface preferred, unless the specific advantages of Seriplex are compelling for a specific project.

WorldFIP-I/O

WorldFIP-I/O is an international standard sensor network and
is supported by a few control equipment suppliers including
Schneider Electric and Alstom. WorldFIP-I/O is a proper subset of WorldFIP, an international standard fieldbus. Alstom
products are called MicroFIP and deliver high-speed synchronous transfers of data for distributed real-time control. World-
FIP-I/O is used as a sensor network by ENEL, the electric
power utility of Italy, CERN, the large nuclear research institute in Switzerland, and on the French TGV high-speed train.

Although WorldFIP-I/O is technically a great sensor network,
it has not been accepted as such by enough users to make it
successful. Sharing WorldFIP-I/O implementation with World-
FIP fieldbus was designed to increase the volume of usage and
lower the cost, but WorldFIP usage in all forms remains too
small to provide that low-cost advantage. WorldFIP-I/O is also
a European standard sensor network, sharing those honors
with AS-interface, Profibus-DP (reduced profile for use as a
sensor network), and Interbus-Loop.

CAN

CAN (Control Area Network) was originally developed to replace wiring harnesses for distributed I/O in automobiles and trucks. While it has been used for this purpose to some degree, the major automobile manufacturers have yet to deploy it broadly. CAN has been used as the technical basis of DeviceNet, SDS, CAN Kingdom, and CAN Open networks, but these have not concentrated on cost reduction of the wiring. The CAN chip contains a small programmable processor, and these networks have concentrated upon providing a highly functional user layer for discrete I/O, much like the fieldbus networks have done for analog I/O. For this reason, these networks are classified as fieldbuses. However, it is expected that the number of major automotive manufacturers deploying CAN to reduce the cost and lighten the automobile by creating an under-the-hood sensor network will greatly expand over the next few years.

Both DeviceNet and SDS have been embedded into sensors and actuators. SDS has remained somewhat proprietary to its inventor, Honeywell Sensing and Control, although the specification is fully open and available at the Honeywell web site. Honeywell has implemented SDS throughout its broad line of sensors and actuators as *smart devices*, but they do not offer an SDS termination block similar to AS-interface, Seriplex, or even DeviceNet. Similarly, the DeviceNet specification is open and available from ODVA (Open DeviceNet Vendors Association) and has been embedded into sensors and actuators by many ODVA member companies, including Rockwell and Cutler-Hammer.

DeviceNet is the only version of CAN to be offered as a sensor network. I/O modules similar in function to those of AS-interface and Seriplex are available to terminate conventional sensors and actuators and communicate on the DeviceNet network. Although the higher level application layer, called CIP (Common Industrial Protocol), can be used with DeviceNet as a fieldbus, it need not be used, making DeviceNet no more complex as a sensor network than AS-interface or Seriplex. Cost of a DeviceNet node may be a factor for any applica-

tion, but using DeviceNet as a sensor network will yield the
same wiring savings as AS-interface or Seriplex.

DeviceNet is offered with 4-conductor round (not illustrated)
and flat cables, illustrated in Figure 4-10, delivering electrical
power to the nodes and devices. Power is delivered separately
from data in both cables. The DeviceNet specification does not
call for a wiring standard between the device and the
DeviceNet interface module. Most DeviceNet interface
modules use compression screw terminals for bare wire, but
the M12 round connector is often used to connect devices to the
DeviceNet interface module. Use of the flat cable with
DeviceNet interface modules as illustrated in Figure 4-11
provides for a low-cost installation very close to AS-interface in
cost. DeviceNet is also available with higher density interfaces
using bare-wire terminations. The cost of a DeviceNet network
can be almost as inexpensive as AS-interface for most
applications.

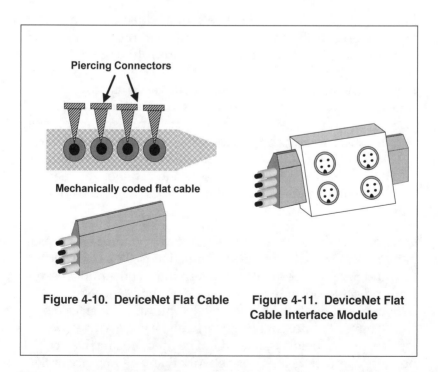

Piercing Connectors

Mechanically coded flat cable

Figure 4-10. DeviceNet Flat Cable   **Figure 4-11. DeviceNet Flat
                                                  Cable Interface Module**

Compared with the simpler sensor networks, CAN-based networks claim to have a more secure communications environment provided by the protocol. The cost of this security may be substantial, but only you can determine its value. Using CAN-based protocols offers the advantages of signal processing at the field device to off-load high-speed operations from the PLC, a typical advantage of any fieldbus, but only SDS products offer any functionality using the processing power of the CAN chip.

LonWorks/LonTalk

LonWorks has been used as a sensor network primarily in building automation, security systems, and HVAC (heating ventilating and air conditioning) applications. It also features a chip with a programmable microprocessor. Rather than concentrating only on cost reduction of the wiring, LonWorks has concentrated on distributing some of the control processing to each network node as a fieldbus. However, LonWorks uses low-cost twisted-pair wiring in multidrop configuration, and can be used as a sensor network. LonTalk is the name of the protocol used for the LonWorks network. LonMark is the vendor-support organization that validates products conforming to the LonTalk protocol in the same way as the Fieldbus Foundation supports Foundation Fieldbus. There are many products displaying the LonMark symbol including low-cost device interfaces for discrete I/O.

The LonWorks program has been substantially ignored by the industrial automation community. However, LonTalk is the standard network protocol for building automation systems including HVAC, security, elevator and escalator controls, fire safety, and intrusion detection and alarms. There are thousands of products from major suppliers available for the LonTalk network, which is an international standard. Yet, there are few industrial interfaces for LonTalk on PLCs or DCSs, and those suppliers that have a LonTalk interface do not make LonWorks compatible I/O. This makes it difficult to configure a LonWorks system for any industrial project. In spite of these difficulties, LonWorks has been used in many water- and waste-treatment automation projects because it offers low-cost, adequate performance, and can be applied in large physical areas. Of particular interest is the variety of physical media over which LonWorks can be implemented including twisted-pair copper wire, power line carrier, and wireless.

Review Questions, Sensor Networks

1. Why should you consider use of a sensor network?

2. When would it not make sense to use a sensor network?

3. What is the most important factor in selecting a particular sensor network?

4.2   Fieldbus Applications

When fieldbuses are used in the factory, the purpose has been to reduce installation cost by moving the I/O interface from the PLC to a remote I/O unit mounted close to the machine on the factory floor. Factory automation fieldbuses are fast and deterministic. In this case, deterministic means that the maximum worst-case time to obtain data across the fieldbus can be accurately predicted and is not subject to chance.

In the past, there was a distinction between fieldbuses and devicebuses, a name that was assigned to bus technology intended for discrete automation. However, as a byproduct of the Fieldbus Wars of the 1990s, several devicebuses (Profibus and Interbus) were included in the IEC 61158 fieldbus standard. Since most of the former devicebuses are capable of transporting scalar data, the distinction is no longer relevant. All of the network technology formerly called devicebus is included in this book as a fieldbus.

All of the fieldbuses used for factory automation allow the remote device or I/O unit to be intelligent and to execute software. All of these fieldbuses also can be used to interconnect PLCs and PCs into a network to share information. However, there is no specific software to allow control logic to bridge across the network. This means that logic in one PLC cannot link directly with logic in another PLC on the same network. It is possible to accomplish these same goals, obtain the status of an I/O point or an intermediate variable across the network to use in logic, but the timing of network access is not in synchronization with the timing of PLC logic and is usually not deterministic.

Fieldbuses used for process control were explicitly created to link smart field instruments to each other and to higher level control systems. While the speed requirements for process control are much less than for factory automation or material handling, within that speed the requirements are no less for determinism. In fact, process control adds an additional requirement for very tight time synchronization. Additionally, the amount of data exchanged for process control is far larger than for factory automation or material handling, and almost always involves several floating point numbers as well as several discrete status bits.

For many applications, process control fieldbuses are also required to conduct electrical power to field instruments over the same wires used to communicate data. In keeping with modern safety requirements in the chemical and petroleum industries, process control fieldbuses must prevent ignition of flammable gases by conforming to the requirements for intrinsic safety. These application requirements stem from the legacy of process control field communications, analog 4-20 mA that met these same requirements.

4.2.1 Factory Automation Fieldbus Applications

The original application for which factory automation fieldbuses were designed is to connect remote I/O termination units to the PLC, as illustrated in Figure 4-12. The original remote I/O unit was a small equipment rack to contain the PLC's I/O cards, but modern remote I/O, often called *block I/O*, often appears as an electrical termination block for

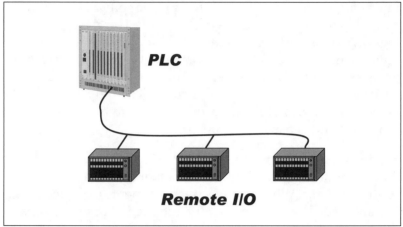

Figure 4-12. PLC with Remote I/O

connecting sensors and actuators. The signal-processing intelligence of the fieldbus is contained in the block.

There is an overlap between factory automation fieldbuses and sensor networks. Sensor networks cannot perform signal processing or simple control operations in the field but require the PLC to perform all logic. The processors used to implement fieldbuses are perfectly capable of performing signal processing for discrete I/O and even simple logical controls, but control operations are rarely done at the fieldbus level.

It is possible to install intelligence into binary 2-state sensors and actuators with fieldbus communications. Two of the CAN-based fieldbus protocols, DeviceNet and SDS, have defined a number of functions only possible when there is a microprocessor operating in the field device with a direct connection to the sensor or actuator. High-speed filtering and signal-processing functions supported by discrete automation fieldbuses become very practical with the close connection to the sensor, whereas remote connection to a PLC would require impractical high-speed polling. All of these functions are also possible in a fieldbus interface directly coupled to several sensors and actuators, except for those functions that require a mechanical connection.

A few of the switch manufacturers make products such as limit switches, proximity switches, photoelectric detectors, and some other devices with the fieldbus network interface built into the device itself. Except for a few unusual applications, these products have been too expensive for general use since the single-point fieldbus interface is more costly than the same connection without the fieldbus network interface. Where sensors are not clustered, use of a device with a built-in fieldbus interface may lower the cost of installation since a separate fieldbus interface would not be required. However, most of the signal processing is also available through the separate fieldbus interface block. The attraction for these devices with a built-in fieldbus interface is their extensive signal processing and tuning keyed to the sensor. Table 4-1 illustrates some of the built-in functionality useful for a limit switch or pushbutton through a fieldbus interface.

Table 4-1. Example: Limit Switch or Pushbutton Functions

Function	Description
De-bouncing	Suppresses the intermediate state changes when a contact opens or closes.
Time delay	Imposes a configured time delay on OPEN or CLOSE or both before reporting the state change.
Diagnostic test	Reports result of internal diagnostic tests such as contact resistance.
Pulse count	Counts pulses from each OPEN/CLOSE cycle; may reset to zero each read cycle or not
Pulse duration	Measures time between the last OPEN/CLOSE cycle.
Switch cycles	Number of OPEN/CLOSE cycles in lifetime of switch.
NO/NC	Configure switch as Normally Open or Normally Closed.
Deadtime	Configure a time during which switching action is not reported.

Similarly, Table 4-2 illustrates some actuator functionality that might be useful for a solenoid valve or electrical switch with a fieldbus interface. Electrical output switches are often used to

operate lamps or activate electrical motor starters or other
electrical power contactors. Few products are available in this
category.

Table 4-2. Example: Solenoid Valve or Electronic Switch Functions

Function	Description
Time delay	Imposes a configured time delay on OPEN or CLOSE or both
Actual output status	Mechanical or electrical feedback digital input of output state
Flash	Operate output OPEN/CLOSE at configured cycle time
Pulse count output	Cycle output OPEN/CLOSE/OPEN for a number of cycles at a configured cycle time
Pulse duration output	Cycle output OPEN/CLOSE/OPEN once for a time duration

Fieldbus functions for discrete automation are explicitly
designed to remove some of the higher speed operations from
the PLC. The effect would be to reduce the criticality of scan
speed on PLC operation and to remove some of the more repetitive and often forgotten functions from the PLC's logic. For
example, PLC programmers often forget to program contact
de-bounce into PLC logic unless the bouncing makes process/
machine operation less reliable. Then de-bounce is often implemented with a simple time-delay function. Microprocessors
located in the fieldbus interface often have more sophisticated
algorithms for contact de-bounce since they are local to the
device and see each state change.

Discrete automation fieldbus functions are not often used in
the configuration or programming of logical controls, even
when fieldbuses are used. The discrete automation fieldbus
functionality was developed by PLC suppliers, but they soon
recognized that this use of fieldbus could reduce the need for
high-end PLCs and profitable I/O interfaces. Sensors and actuators with built-in fieldbus interfaces and functionality have
not been popular, due to the very high incremental price of the
fieldbus interface dedicated to that single device. Only when

the fieldbus functionality is contained in a shared device using conventional sensors and actuators has there been marketplace success. Even then, the functionality of the fieldbus interface is not often used.

Programming for discrete automation has traditionally used Relay Ladder Logic (RLL) or Structured Text (ST) methods. While these are equivalent, they are very different in context and structure. They are, however, identical in model: the PLC does all of the control functions on a non-synchronized cyclic schedule. RLL or ST do not enable programming the functions of a remote device, such as those on a fieldbus. It becomes too easy to simply use the unprocessed input and use a timer function in the RLL or ST environment.

Motor Control Center (MCC)

The first example application is one in which intelligent fieldbus devices are used for electric motor starting control independent of the PLC. This example was selected to illustrate the potential of remote network intelligence. There is more to controlling a constant speed AC electric motor than an ON/OFF switch. Before fieldbuses were used, a motor starter or contactor contained in a motor control center (MCC) simply had an ON/OFF input switch and solenoid operated contacts for the high-voltage electrical power. The information supplied in the form of discrete signals was typically motor running, motor stopped, and overload trip.

A typical fieldbus connected MCC has the motor ON/OFF input but provides far more information on the electric motor being controlled. The map for the information contained in the MCC is contained in a file delivered by the MCC supplier or, in some cases, is a standard for MCCs described by one of the fieldbus trade associations. For MCCs supporting DeviceNet, the file is called an EDS (electronic data sheet), while for Profibus-DP, the file is called a GSD (Gerätestammdaten: equipment master data). Both the EDS and GSD are highly formatted text files that map the structure of the registers contained in the MCC down to the bit level. Neither EDS nor GSD is written in XML format at this writing.

Older communications systems such as Modbus, Modbus+, and most of the proprietary protocols such as DH+ only access registers of data in the remote device. This requires that the programmer refer to the device documentation to determine where the desired data is located in the set of registers, then program the extraction of that data through Boolean masking and shifting register data. EDS and GSD allow mapping of basic data to bits in a register for all similar devices and further allow device manufacturers to extend this information through the use of additional bits. Instead of knowing which bits of which register can supply the desired data, EDS and GSD allow direct access to the desired register or set of registers, but the bit assignments must still be retrieved using information in the data file. The real advantage is that many software packages are capable of extracting the desired data if they are EDS or GSD enabled.

Perhaps by the time you are reading this, OPC/DX will be in common use. OPC/DX provides for a more symbolic access to parametric data from automation devices. The OPC/DX server takes the responsibility of transferring the desired data from the device based on specification of the name of the parameter.

Figure 4-13 illustrates some of the data flow between an intelligent motor control center (MCC) and a PLC using DeviceNet and EDS. While simple devices will always have an electrical signal for turning the motor ON or OFF and will supply a Running indication, only an intelligent MCC can supply data such as average current and diagnostics, including warnings preceding overload protective relay trip. Often intelligent MCCs will supply many other items of operational data. It is also possible that limited sequence operations can be done in intelligent motor starters, but VSDs (variable speed drives) can be programmed to do such operations as starting under full load, speed ramping, and motor reversing.

Machine Safety System

A second example is a safety interlock system typically used for moving machinery that might be hazardous to human life. Machines such as stamping presses require human operators to

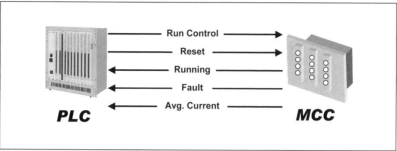

Figure 4-13. Communications Between PLC and Motor Control Center

set up the work and tools, but the operator must be careful to stand clear of the machine as it does its work. Safety floor mats are used to detect if an operator is standing in a hazardous location and also to be sure that the machine operator is standing at a safe location. Light curtains are used to be sure that the operator does not have arms or legs in the hazardous area. Dead-man switches are used to prevent machine operation unless the operator is in a safe place. Figure 4-14 illustrates this machine safety system.

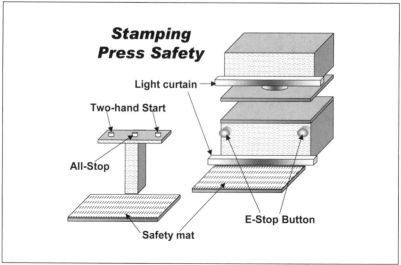

Figure 4-14. Machine Safety System

The illustrated safety system for the stamping press is quite complex, with dual safety mats, a light curtain, dual start push-buttons, two emergency-stop buttons, and an all-stop button. The plan is that the operator positions the work on the press table, then goes back to the operation station where he must press both start switches at the same time. The press will cycle to stamp out the part only if the light curtain indicates "all clear," the floor mat at the machine is open (no person on the mat), and the floor mat at the operation station is closed (a person is on the mat). The machine will stop its cycle if the all-stop button is pressed and will not restart unless the start conditions are all true. When either of the emergency-stop buttons is pressed, the machine will immediately halt operation, and the press will be full opened. To prevent cheating, the logic for the start buttons can require that the two buttons be pressed within 100 ms to prevent the operator from jamming one button down so that he can have one-handed operation.

Previous safety system practice required that all of these switches be directly wired to a dedicated safety processor or a master safety relay. Modern safety practice allows the machine control PLC to be used if it meets the specifications for the appropriate safety level. The many switches can also be wired using fieldbus technology provided that the bus technology detects any bus failure or any safety device failure and then will cause an emergency-stop trip. Fieldbus redundancy is not necessary, but if used appropriately, it can make the system more reliable by making bus failure less likely. Both hardware and software/protocol methods can be used to detect failure of the bus and the safety devices. Safety versions of DeviceNet, Foundation Fieldbus, and Profibus exist for this purpose.

If Foundation Fieldbus is used for the safety and operational I/O, local control becomes possible. Local control is allowed but not supported on DeviceNet, Profibus, Interbus, Modbus, or SDS. Local control is supported on LonWorks, but programming is required. A Foundation Fieldbus network node with a Flexible Function Block can run logic programs directly using the local I/O, but is not often used in factory automation applications. Local logic control can also be done with a small dedicated PLC located at the machine, in which case a fieldbus may not be required.

Review Questions, Discrete Automation Fieldbuses

1. What is the main reason to use a fieldbus?

2. Networks used for control must be deterministic. What does *deterministic* mean?

3. What are some of the advantages and benefits of using smart discrete I/O?

4.2.2 Process Control Fieldbus Applications

Typical process control operations are slower than those for factory automation applications, allowing slower network speeds but often requiring larger data transfers. While factory automation networks must be fast, there are few demands for tight time synchronization between network nodes. Process control, while not as demanding on network speed, does demand tight time synchronization between nodes forming a cascade control loop.

> The original DCS was conceived to have many controllers with all control loops *functionally distributed* among them. It was also an objective to *geographically distribute* controllers to save on wiring cost. We continue to use the term "distributed" control to mean implementation on a DCS.

Foundation Fieldbus is designed to enable time-critical closed loop control data to be exchanged on its fieldbus network. The architecture of Foundation Fieldbus is called *Field Control*, or the seamless construction of a control loop consisting of a highly interconnected set of function blocks to perform the closed loop control function with, or without, the participation of a host controller. Field Control is generally considered by those who favor it as being more reliable, more accurate, less expensive, and more responsive than distributed control. The reasons for these claims are itemized in Table 4-3, which also provides some of the benefits frequently cited for Field Control.

Table 4-3. Field Control Benefits

Claim	Rationale	Benefit
Reliability	Field instrumentation is constructed for extremely high reliability	No loss of control
Accuracy	More frequent control is possible with a dedicated processor	Deviation from setpoint, process variance is reduced
Lower cost	The number of host controllers is reduced; incremental cost of control in field devices is small	Less expensive control systems
Responsive	Final control in a control valve positioner can reduce deviation from setpoint	Less process variation enabling tighter control

There is much debate on the virtues of Field Control vs. Distributed Control, most of it meaning nothing if the system is well designed. Most DCSs have implemented control with all cascades limited to a single controller, meaning that cascade setpoints are not communicated to the control level network. Most DCSs allow many controllers to exist and share information on the control level network as long as it is not time-critical closed loop control information.

Many of the benefits of Field Control are based on *better control,* which is not often discussed in textbooks on process control. Better control means that the normal oscillation of a closed control loop about the setpoint is smaller. Generally, removal of deadtime and hysteresis from the control loop will result in control with fewer and smaller deviations from setpoint – *if* the control loop is correctly tuned. Field Control can remove deadtime and hysteresis from the loop.

Conversely, proponents of Distributed Control, as in the classical DCS, have their own set of valid claims as well. Actually, there are two parts to the control story: signal processing and closed loop control. Both groups favor signal processing in the field device closest to the actual sensor. Signal processing consists of conversion of the raw data from the sensor to engineering units after removing the effects of noise and thermal drift. This could be done in the controller, but more and higher frequency data collection would be required. Doing the signal processing close to the sensor eliminates the need to send high frequency volumes of data. Additionally, alarm and limit testing can also be provided in the field device in either architecture.

Distributed Control proponents claim that it is less costly and simpler to do the control function in a reliable device specially built for this purpose, and share the cost among many control loops. With the cost of control room electronics constantly in decline, there is some merit to this claim. Simplicity of doing all control processing in a few multifunction controllers is a given – it is far simpler than Field Control. Additionally, there are many control loops that are far too complex for Field Control.

For most processes, a combination of Field Control and Distributed Control is probably the most economic, responsive, accurate, and reliable. Foundation Fieldbus enables this type of mix for all DCS architectures. Profibus-PA enables field signal processing, but does not enable Field Control. WorldFIP shares a common architecture with Foundation Fieldbus, but is not often used in process control applications. HART does not permit Field Control, but is fully capable of field signal processing, although is not often used for this purpose.

Review Questions, Process Control Fieldbuses

1. What are the differences between a process control fieldbus and one used for discrete automation?

2. What is the difference between *Distributed Control* and *Field Control*?

3. What are the benefits of Field Control?

4. Which fieldbus should be used for Field Control?

Distributed Process Control

A classic control loop in control of chemical, petroleum, and petrochemical processes is the BTU (British Thermal Unit) controller for the bottom of a distillation column. Distillation is a multivariable process with many interacting processes. In modern control designs, the entire distillation column would be controlled using model predictive control (MPC), which usually requires significant processing power. MPC removes deadtime, hysteresis, and interaction among control loops by using a dynamic process model tuned for the column, production rate, and product mix. Due to the complexity of MPC, it usually operates as a high-level controller, making it too slow for control of most physical variables such as the level in the bottom of a distillation column.

The controls illustrated in Figure 4-15 consist of physical property measurements of level, steam temperature, steam pressure, volumetric flow of steam, and flow of distillation bottoms. The controlled variables are the flow rate of bottoms and the flow rate of steam. The objective is to hold the level at the bottom of the distillation column constant, to balance the flow of bottoms consistent with the column feed stream, and to add enough heat through steam to provide the necessary internal vapor rate for distillation. Experience has shown that control of level using the bottoms flow rate is too unstable, so control of level is assigned to the setpoint of the steam flow rate. In this way, the bottoms draw rate can be used to control the column's internal vapor flow. However, even controlling steam flow rate can be less than stable when there are fluctuations in steam quality. Therefore, experience teaches us to control the flow of BTUs (or Calories) through the steam flow. This requires computation or table lookup of the BTUs per pound (or Calories per kilogram) of steam at the measured temperature and pressure of the steam.

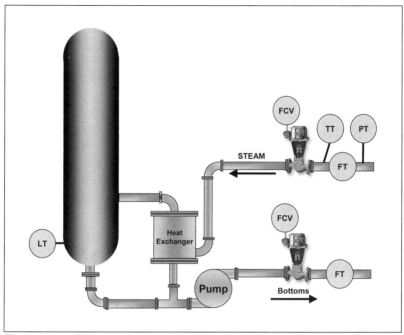

Figure 4-15. Bottom of Distillation Column, BTU Control

Implementation of the BTU controller requires solving a complex nonlinear equation, or looking up the steam heat content (enthalpy) in Steam Tables[1]. At present, this is considered out of scope for computations in a field device and is more appropriate for a DCS using a multifunction controller. Additionally, the bottoms control is only one part of the control of a distillation column involving many interdependent and interactive control loops. For example, the primary controlled variable in the distillation column is the internal reflux ratio, which cannot be measured but must be computed. The most frequent implementation of modern distillation column control is to use model predictive control for all control loops except for the final control elements such as steam flow rate and bottoms flow rate. These two control loops must be done in the DCS controller if HART or Profibus-PA is used as the network, but may be done in the DCS controller or in the field instrumenta-

---

1. *Thermodynamic Properties of Steam*, American Institute of Mechanical Engineers, 1967

tion using Foundation Fieldbus function blocks if Foundation Fieldbus is used as the network.

Foundation Fieldbus H1 or Profibus-PA instruments may be wired in multidrop or daisy-chain topology, but for reduced installation and maintenance reasons, are usually wired in star topology. In this topology, each instrument is wired to the fieldbus segment termination in a common junction box. Each field instrument will have a dedicated wire pair from the instrument to the junction box where each fieldbus segment is terminated as illustrated in Figure 4-16. A wire pair from the junction box to the controller is necessary for each fieldbus segment. Control level networks such as Foundation Fieldbus HSE or Profibus-DP can be used to reduce these home-run wires. Conventional analog control, including HART, always requires a dedicated wire pair from the field instrument to the controller.

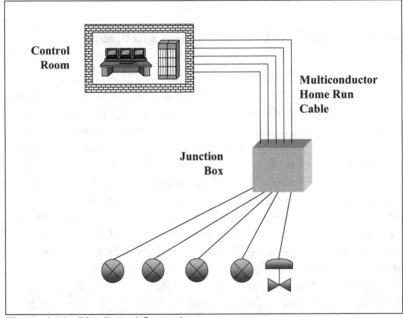

Figure 4-16. Distributed Control

Field Control

There are many types of boilers in commerce and industry. Power boilers are used to make steam at high pressure for operation of turbogenerators. Heating furnaces are boilers used to make steam and/or hot water for heating buildings. Recovery boilers are used in the pulping industry to process waste chemicals and to make steam for pulping and paper-making. All boilers burn fuel with air and produce waste gases and lots of heat. Power boilers are usually equipped with extensive control systems to keep them operating at peak efficiency, extracting as much energy from the fuel as possible. Recovery boilers are considered as power boilers, but are also controlled to minimize the use of purchased fuel as they burn waste products.

Modern power boilers control excess oxygen by regulation of the inlet air damper position using measurements of carbon monoxide (CO) and oxygen in the flue gas. If there is more than a trace of CO, the air damper must be opened to supply more oxygen. If there is excess oxygen in the flue gas, too much air is being used, and the air damper must be closed. Generally, the fuel flow rate is controlled by the pressure of the steam being produced by the boiler, since the steam is used continuously by turbogenerators as they in turn respond to the demand for electricity. Figure 4-17 illustrates a simplified control system for a power boiler.

In this example, designed around the use of Foundation Fieldbus function blocks in the field instrumentation, controls for the feedwater flow rate, the fuel flow rate, and the air-fuel mixture are shown. Both CO and excess oxygen (O_2) are sensed in the exhaust gas stack from the boiler. Each measurement is fed to a PID loop controller; stack gas conditions will indicate that there is excess O_2 or the presence of CO in the exhaust gas, but never both at the same time. The outputs from both controllers are sent to a Control Selector (CS) function block that selects the controller with the active output; the other one can only output zero. When the O_2 controller is selected, the PID output is used to close the air damper, while the CO controller output would open the air damper. The air damper positioning is con-

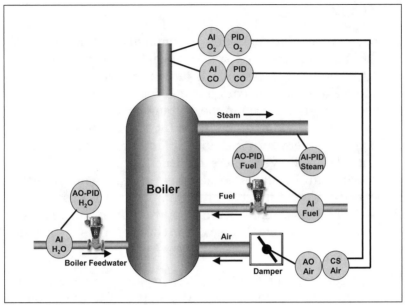

Figure 4-17. Simplified Boiler Controls

trolled by the AO (Analog Output) function block. The Control Selector function block always sends data back to both PID controllers indicating if they are in control, and if not, the reset remainder is cleared to prevent open-loop windup. The fuel feed controller setpoint, in this example, is cascaded from the steam header pressure controller, applying more fuel as the pressure is reduced by increased steam demand. In this example, the boiler feedwater flow rate setpoint can be based on the level in the boiler drum, ratioed to the fuel setpoint, or cascaded to a measurement of steam flow; for simplicity, none of these are shown. This complex control scheme can be implemented entirely within the distributed field instruments, but only by using Foundation Fieldbus as the network. It may also be implemented in the DCS controller.

Figure 4-18 shows the wiring advantage of using a Field Control system, with its two-level network structure. All of the fieldbus instruments are connected individually to a field junction box where a linking device multiplexes all fieldbus segments into a control level network, such as Foundation

Fieldbus HSE or Profibus-DP. When connected more traditionally, as was illustrated in Figure 4-16, not only is a wire pair
required for the home-run cable segment, but fieldbus terminations are required for each fieldbus network segment. The combination of reduced wire pairs, fewer bus terminations, and the
accompanying reduction in installation labor makes the two-
level fieldbus approach less costly to install and maintain.

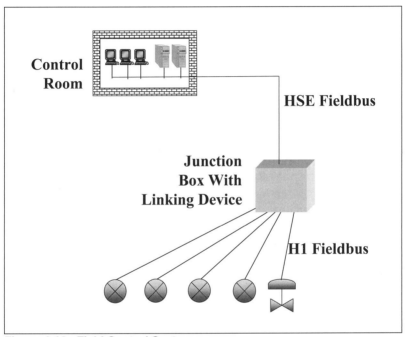

Figure 4-18. Field Control System

4.2.3 Building Automation Fieldbus Applications

Building automation is a blend of discrete and continuous
feedback loop control depending upon the building process
being controlled. HVAC (heating, ventilating, and air-condi-
tioning) is very much like process control with temperature
and humidity as the controlled variables. The manipulated
variables such as air flow, heating coils, steam flow, and com-
pressors tend to be controlled with ON/OFF states and time
delays, rather than modulated control elements. One of the pri-

mary objectives of a building automation system is to provide
the needed services at the minimum investment. Control accuracy is rarely a requirement. Since manipulated devices tend to
be ON/OFF, controllers are typically optimized to create on/
off outputs.

HVAC Building Control System

Modern buildings are engineered with all-digital temperature
and humidity controls. Since people tend to prefer setting their
own temperature setpoints, the controllers tend to all be distributed to the areas under local control. In addition, there are
typically centralized controls to provide temperature overrides
for meeting rooms and for nights and weekends. Since the output of a temperature controller is an ON/OFF signal, some
variability around the setpoint is to be expected. Additionally,
most heating and cooling processes retain their heating or cooling ability for a short time period after being turned off. The
mechanism provided to compensate for the retained heating or
cooling is called an *anticipator*. In older analog controls, the
anticipator was a small heating coil in the controller itself. In
modern digital controls, the anticipator is a time-lead function
block.

The standard network for HVAC is called BACnet, which is
actually an application programming interface (API) for the
use of any suitable network. While the BACnet standard offers
a lower level network based on the use of EIA-485, it has been
more often implemented on LonWorks and Profibus, both of
which are approved technologies for HVAC.

Similar to older analog HVAC systems, a digital building control system uses local ON/OFF temperature and humidity controllers in each area needing temperature and humidity
control. The setpoint for these controllers may be manually set
by occupants for private offices, may be hidden for group
areas, and may be remotely set for some areas. Buildings often
have a supervisory console for HVAC to allow remote setting
of setpoints as well as night, weekend, and holiday setbacks to
conserve energy.

The LonMark Interoperability Association (LIA) is the organization responsible for conformance testing of devices operating on a LonWorks network. Additionally, LIA develops and publishes formal object definitions called LonMark Functional Profiles for the intelligent devices on the LonWorks network. The LonMark Profile defines the external interface to the network object in much the same way as a Profibus profile. The algorithm of the controller, usually ON/OFF or PWM (pulse width modulation) is not defined, to allow this definition by the supplier.

Elevator Control System

An elevator is a complex device requiring a very simplified user interface for the general public, and extensive safety systems to protect human life. Most systems involve the coordination of several elevators to avoid unnecessary user service delays. The controller for elevator movement cannot be distributed; by law it must be stationary, usually at the top of the building. However, many of the safety and local function operations are distributed and ride on the elevator car itself. Use of a fieldbus for control of elevators eliminates a large number of wires that previously were required to connect the moving car and all of the pushbuttons and lights on every floor to the elevator controller at the top of the building. Figure 4-19 illustrates a typical single elevator control system.

The elevator car illustrated in Figure 4-19 shows many of the devices located on the car and on each floor as well. Since the elevator is suspended on long steel cables driven by a variable speed reversible electric motor at the top of the building, and cables stretch, precision alignment at each floor requires the use of a floor sensor mounted on the car that reads a location device in the elevator shaft at the floor level. Once the car has stopped at the floor, a floor lock is engaged to prevent movement of the car while people enter or exit. Once the car has arrived at the desired floor, the inner doors are opened simultaneously with the outer doors located at the floor landing.

People entering the car may push new floor buttons. When they do, a signal is sent to the elevator controller that in turn

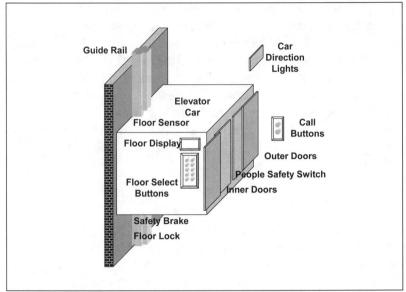

Figure 4-19. Elevator Instrumentation and Display

sends a signal back to the car to illuminate that button on all panels in the car. This illumination must come from the controller and by code, must not be generated locally. Usage testing has confirmed the need for a separate illuminated button for each floor and not a numeric keypad. After a suitable waiting period, or when a passenger presses the close door button, and the open door button is not being depressed, the doors will close followed by disengagement of the floor lock. Then the elevator controller will signal the variable speed drive to move the car in the proper direction by ramping to the speed appropriate for the number of floors being traversed. An indicating panel inside the car shows the current floor location of the car along with indications of the direction of movement.

On each floor of the building, there is a pushbutton panel with UP and DOWN buttons. Usually, there is an indicator showing the direction in which the elevator is moving and often the floor where the elevator is located.

The optical safety systems in the car prevent doors from closing if people are detected in the doors' path. Usually there is a multilevel photoelectric detector and switches located at the leading edge of the doors. If people are detected, the doors are opened and a wait cycle is started before trying again. If something happens to the drive mechanism and the car begins falling too fast, a friction brake on the car is activated to stop the car. If the cable breaks and the car begins a free-fall, an inertial safety brake is automatically engaged.

Of all the actions described above, only the inertial safety brake is mechanical. All of the other items are driven by logical controls in the centralized elevator controller. The speed ramping between floors is driven by a speed control algorithm, often using the actual weight of the car determined from actual car movement and the current draw on the variable speed drive motor.

Use of a LonWorks fieldbus has dramatically reduced the cost of elevator installation and long-term maintenance, while increasing reliability. Each of the pushbutton panels in the cars is connected to the topworks controller using redundant twisted-pair cabling. Multiplexing of the individual pushbuttons and illumination lamps is done locally on each panel. All pushbutton and indicator panels are connected to the same fieldbus cable. Previously, a wire pair was required to connect each pushbutton and lamp to a termination on the controller, a long run of cable from the top of the building to the bottom that was constantly flexing as the car moved. The redundant fieldbus cables have far fewer connections and wires and only connect panels. The wiring of the panels is not done in the field but in the shop where it can be completely tested. The labor formerly required to validate each connection is eliminated. Intelligence at each panel allows each I/O point to be uniquely identified.

4.3   Control Level Network Applications

Control level networks are currently differentiated from fieldbus networks. Control networks are expected to have higher bandwidth (more throughput) and do not need to deliver electrical power or be intrinsically safe. The intended use of control networks is to link controllers to host systems and to each other. The other application for control networks is to join fieldbus segments together and to host systems. While a number of unique physical layers have been developed for control networks, the clear trend is for a control network to be implemented using commercial off-the-shelf Ethernet components, and common Internet protocols such as TCP/IP/UDP. The objective is to take advantage of the low cost and high performance of commercial networks, even when industrialized components need to be used. Generally, the application layers running above control networks are the same as those used for the corresponding fieldbus segment to which they connect.

In some cases, control networks are being extended to field devices such as remote I/O, variable speed drives, and field mounted PLCs. While there are functional differences between control networks and fieldbuses, it can be expected that the lower cost Ethernet-based control networks will increasingly be used in fieldbus applications when the features of a specific fieldbus, such as intrinsic safety or electric power delivery to the field device, are not needed.

Since most of the current control level networks such as Foundation Fieldbus HSE, EtherNet/IP, Modbus/TCP, PROFInet, and iDA are all based on use of 10/100BaseTx Ethernet with EIA Category 5E cable, they all benefit from progress in the definition of the Power over Ethernet standard, IEEE 802.3af. This standard defines the way to transmit 48 V DC power on the signal pairs or the spare pairs of Cat 5E cable. The standard allows devices on the cable to take power from the cable or not. It also allows the use of devices, built before this standard was ratified, that know nothing about power over Ethernet. However, there is no provision for industrial users needing to limit power or voltage for intrinsic safety reasons.

The oldest of the control level networks is Modbus. The original Modbus was defined as RS-485 multidrop on a single twisted-pair cable using FSK modulation at 9600 baud. Later variations increased the speed and specified higher levels of error correction. However, the real contribution of Modbus was the application commands that standardized the functions of reading and writing registers, starting and stopping PLCs, forcing outputs for testing, and many other data manipulations driven from a host computer. The Modbus specification has long been made public and is in wide use by many devices including interfaces to all major DCSs and most PLCs. A variation of Modbus called JBus also exists. The only difference between Modbus and JBus is the numbering of the registers – Modbus starts with zero, JBus starts with 1 – commands are identical.

While the Modbus network in all of its forms remains very popular, the trend is clearly to migrate to Modbus/TCP for control level network applications. Modbus/TCP allows any host or PLC to exchange data with any other node on the network and is not confined to the single master requirement of Modbus. Modbus/TCP opens the channel to allowing PLCs to directly exchange data with other PLCs without relaying through a host system. Ethernet provides a low-cost, high-speed connection with many commercial options.

ControlNet was developed long after Modbus, to make open the interface to Allen-Bradley PLCs that previously had used a proprietary network called DH (Data Highway). Not only is ControlNet an open specification, but it is faster and more secure and provides for time synchronous data transfers. While ControlNet is still quite popular with Allen-Bradley PLCs, the newer EtherNet/IP network is faster and generally less expensive for the same number of devices. Additionally, both networks use the same user layer called CIP (Control and Information Protocol) also shared with DeviceNet.

Profibus is also used as a control level network. Access to data contained in controllers and other field devices becomes available when using the Profibus FMS Profile, which is contained in both European and international standards. While Profibus can operate over the standard Profibus-DP physical RS-485 network, it may also operate over commercial Ethernet TCP/IP networks, at which time it becomes known as PROFInet. While PROFInet makes Profibus FMS transactions operate on an

Ethernet TCP/IP network, PROFInet opens control level data exchanges to a more modern object technology using OPC.

Foundation Fieldbus HSE was designed to be the control level network for the Foundation Fieldbus network. It too is based on the use of an Ethernet UDP/IP foundation, but the user layer is the same as all other parts of this network. Foundation Fieldbus was designed for use in process control and batch process manufacturing. For this reason, HSE provides strong synchronization for the entire network, and when coupled with its high speed, allows control loop cascades across multiple H1 fieldbus segments. Additionally, HSE provides for network fault tolerance using multilevel redundancy when required for safety and security.

Interfaces for Discrete Automation (iDA) is an emerging network technology that also stresses the importance of tight time synchronization between network nodes. The method used by iDA for time synchronization is called "real-time publish/subscribe (RTPS)" and is very similar to that used in Foundation Fieldbus, but is fast enough for motion control. iDA is object-based using a function block model, also similar to Foundation Fieldbus, but again oriented to discrete automation and motion control. iDA also requires the use of XML encoding in its data transfers. Currently, there are no commercial implementations of iDA.

WorldFIP is also available for use as a control level network but has few applications and few suppliers. The trend for past users of WorldFIP is to migrate to Foundation Fieldbus for process control and to iDA for factory automation applications.

4.3.1 Object Technology

We can no longer think of either control level networks or fieldbuses only in terms of their protocol and wiring. The wiring issue is being resolved as the various control level networks increasingly move to Ethernet and TCP/UDP/IP technology. This also removes many, but not all, of the protocol differentiations. All control level networks are moving to embrace object technology. While there are several object technologies from which to choose, process control, materials handling, and factory automation applications are using OPC (Object Linking and Embedding for Process Control), which is based on the Microsoft Common Object Model (COM).

OPC/DA was originally intended for simple devices such as a PLC, in which both the data and the program could be modeled as sets of registers. OPC/DA enabled the exchange of data with a host computer referencing the registers to be transferred by number, similar to the data transfers of Modbus, Allen-Bradley DH, or Profibus. The difference is that OPC/DA always maps the registers the same, independent of the device, allowing the OPC server to translate between the actual device registers and the OPC virtual registers.

However, OPC/DA only addresses registers of data and the movement of those registers. The data from each PLC is recorded in its own registers, reflecting the way that the I/O is wired to terminations in I/O racks or remote I/O units. The functionality of OPC/DA is only to make the PLC's registers accessible in a uniform way but does not address wiring of the I/O or any control relays. Wiring assignments of I/O points is usually quite different with different PLCs and remote I/O attachments. Generally, the programming of PLCs refers to I/O addresses by the hardware I/O address: rack or remote unit, card, point.

OPC/DX is built on top of OPC/DA, but models the device as a pure object with named attributes. This is highly useful for "smart" process control devices, but many newer discrete sensors and actuators also have addressable attributes ideally suited to OPC/DX.

The effect of using OPC is to hide the automation devices behind the OPC client and server so that the user can treat any OPC compatible device in the same way. For example, it should be possible for an application to be programmed for a control system using one brand of OPC-compliant controllers to run unmodified on an identical plant using a different brand of OPC-compliant controllers. With OPC/DX that is possible, since the symbolic references to data and attributes are independent of the underlying control system.

OPC/UA extends the same objectives of OPC/DX to operate with an object-based data model that fully supports EDDL objects on both Windows and non-Microsoft platforms. For more details, you should refer to the papers, presentations, books, and specifications that can be found at the OPC Foundation website: http://www.opcfoundation.org/.

Review Questions, Control Level Networks

1. What are the main purposes of a Control Level Network?

2. What is the primary technology trend for Control Level Networks?

3. What else distinguishes Control Level Networks from fieldbuses?

4. What is the common high level interface to all Control Level Networks?

4.3.2 Factory Automation Applications

Control level networks are usually used to connect a group of PLCs to the HMI (Human-Machine Interface) stations that serve as the operator control panel for the machine, transfer line, assembly station, or motor control center. In this application, the only control information that passes over the network is the operator commands to turn something on or off. By definition, operator commands are neither time-critical nor imperative. This means that the operator's decision to turn something on or off can be acted upon by the control system in any expedient way and small deviations in time are not important. Also, if the operator command is not recognized by the system, the operator will know and may simply repeat the command (press the button again.)

HMIs are supplied by either the PLC manufacturer or by independent HMI specialty companies. When the HMI and PLC are supplied by the same company, then the control level network selected is usually the network best supported by the manufacturer. If an independent HMI package is used, the network may be any of the control level networks for which the PLC and the HMI manufacturers offer hardware and software support. Generally, any of the open networks can be used with OPC software support. All PLC suppliers offer an OPC server software package, but it usually does not run on the PLC itself but on the PC used to run the HMI software. All HMI software manufacturers supply an OPC client with their software. Table 4-4 shows the control level networks that work with each of the major North American PLC suppliers. Generally, the "Y" means that the PLC supplier fully supports that network; 3rd means that network support is available but from a third-party source.

Table 4-4. Control Level Networks Supported by PLC Suppliers

Control Level Network	Rockwell	Siemens	Schneider	GE Fanuc
ControlNet	Y	N	N	N
EtherNet/IP	Y	N	Y	N
Profibus-DP	3rd	Y	Y	Y
Modbus and Modbus/TCP	3rd	3rd	Y	Y

Y = supported   N = not supported 3rd = third-party software support

PLC system configurations are rapidly changing from large monolithic PLCs with remote I/O units to networks of small PLCs, often distributed close to the machines being controlled. This distributed network configuration is usually less costly to purchase and install than the monolithic PLCs. Figure 4-20 illustrates this modern form of distributed PLC with a combination of HMI stations suitable for machine control applications. Notice that there are a number of HMI panels and displays, some local to the machine, others at a central control room area.

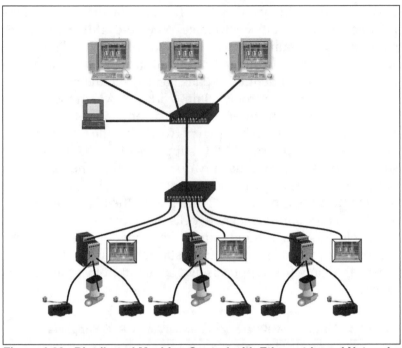

Figure 4-20. Distributed Machine Control with Ethernet-based Networks

The control level network illustrated in Figure 4-20 is based on Ethernet and could be Foundation Fieldbus, EtherNet/IP, iDA, or Modbus/TCP. Each of these networks encodes their data transfers into IP (Internet Protocol) data streams using either TCP (Transport Control Protocol) or UDP (User Datagram Protocol) to send packets on the network. The content of the data packets will be different for each of these control level networks since that is defined by their respective application and user layers.

If Profibus-DP is chosen as the control level network, there can only be a single master station and that must be a PLC or an open control system. Since all HMIs work by polling the PLC for data, it may not be on the same Profibus-DP segment as the one used to connect the remote I/O units to the PLC. Originally, a separate Profibus-DP link was used to connect each HMI to the PLC, but modern configurations typically use Ethernet to link all HMIs to the PLC. Figure 4-21 illustrates this same machine control application with centralized PLC control and remote I/O. All of the HMI stations are connected to common Ethernet network that may be any of the control level networks supported by the PLC supplier as listed in Table 4-4 or a supplier's proprietary network.

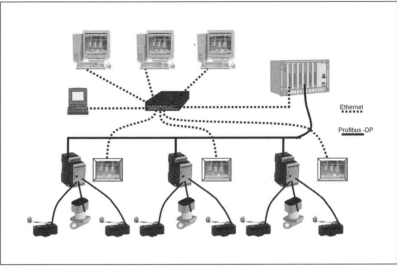

Figure 4-21. Machine Control with Profibus-DP

If ControlNet is chosen as the control level network, the same network can connect all PLCs, remote I/O units, and HMIs. ControlNet does not support network segmentation with switches, and requires careful software configuration to ensure that all communications can occur within the 5 Mbps available bandwidth. However, it is strongly recommended that EtherNet/IP always be used instead of ControlNet on new installations.

4.3.3 Process-Control Applications

The dominant control system for continuous process control is the DCS (Distributed Control System). The *Distributed* part of the title refers to the control function itself, which is usually implemented in several different controllers that may be

located close to the portion of the process under control. A DCS
is defined as an integrated system of controllers, HMI, inter-
connecting network, and application software. While the integrated network of the DCS is a control level network, it has
traditionally been proprietary to the DCS supplier.

An older form of process control uses individual controllers
with built-in simple displays for one or two loops of feedback
control. Often, these are used for smaller plants and to provide
the feedback loop control on batch processes otherwise controlled by a PLC.

Field Control Systems (FCSs) are the emerging form of process
control systems. FCSs are an evolution of DCS in function but a
radical departure in the distribution of control elements. FCS
architecture places the signal processing and control functions
in field instruments instead of in larger multifunction controllers typical of the DCS. Usually the signal-processing functions
are assigned to field transmitters, while feedback loop control
functions are assigned to control valve positioners. An FCS
allows any part of a cascade control loop to be located in a multifunction controller or in an appropriate field instrument as
necessary, and allows relocation of these functions between
devices. Since computing is so widely distributed, a fieldbus is
necessary to connect field instrumentation. Additionally, a control level network is necessary to link the fieldbuses together
when it is necessary to share data between fieldbus segments.
As opposed to DCS, an FCS typically uses standard fieldbuses
and control level networks.

An FCS can use any of the control level networks but typically
will use only those that are compatible with one of the fieldbus
networks. Table 4-5 shows this compatibility:

Table 4-5. Compatibility, Fieldbus and Control Level Network

	Process-Control Fieldbus	
Control Level Network	**Foundation Fieldbus**	**Profibus-PA**
Foundation Fieldbus HSE	Y	N
Profibus-DP	N	Y
ControlNet	Y	N

While any of the fieldbus networks can connect directly to controllers or their local or remote I/O units, several problems
remain that are solved when fieldbus network segments are
linked through a gateway device, often called a network bridge
or linking device. The gateway or linking device usually terminates a number of fieldbuses and connects them to a control
level network. Table 4-6 lists the common problems of using
fieldbuses alone and the benefits provided by using a linking
device or gateway to a control level network:

Table 4-6. Benefits of a Control Level Network

Problem of Fieldbus Networks	Benefits of a Control Level Network
Home-run wiring to controller for every network segment	Higher speed network reduces home-run wiring
Devices on fieldbus segments cannot communicate with devices on other segments	Control level network allows communication between all devices when necessary without using controller software. (Bridging function)
No fault tolerance on longest cable runs	Control level network can provide cable redundancy for fault tolerance

While these benefits would be of value in factory automation
or material handling applications, currently PLC software does
not have a convenient way to distribute control logic or I/O
signal processing among multiple devices. Furthermore, the
fieldbuses used for discrete automation generally have higher
I/O capacity than the fieldbuses for process control, giving the
use of control level networks less of an economic advantage.

Field Control

One of the benefits of using Foundation Fieldbus HSE as the
control level network is that cascade control linkages between
H1 segments can be built with deterministic and synchronous
connections. Before control level networks were available, most
suppliers did not support this type of cascade linkage since
they could not support deterministic bridge timing in software

for the multifunction controllers. However, in Foundation
Fieldbus, the bridge supplied by the Linking Device is part of
the protocol and supports all aspects of network functionality,
including the full network time synchronization required for
cascade control loops.

Figure 4-22 illustrates a portion of a control system in which
the primary and secondary controllers are implemented in
field devices: transmitters and control valve positioners.

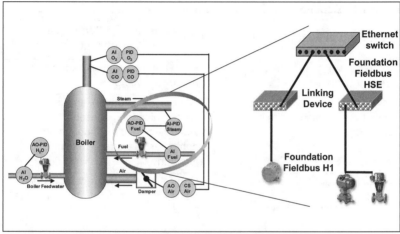

Figure 4-22. Field Control with Foundation Fieldbus HSE

In this example, the fuel-flow transmitter and the fuel-flow
control valve with its positioner are both connected to a single
Foundation Fieldbus H1 segment. The steam flow transmitter
is connected to another H1 segment. The illustration shows the
H1 segments connecting to their own Linking Devices for economy of wiring. Each Linking Device is connected to an Ethernet switch using normal Ethernet wiring conventions for
segmented networks.

Using Foundation Fieldbus function blocks, the transmitters
process the raw signals for steam flow and fuel flow into engineering units values. A PID function block in the fuel-flow control-valve positioner controls the fuel flow by sending its
output to the AO (analog output) function block in the same

positioner. Meanwhile, the steam-flow value is sent to the steam-flow control PID function block in the steam-flow transmitter. The output of the steam-flow controller is sent across the Foundation Fieldbus H1 network to the Linking Device that bridges the signal to the Foundation Fieldbus HSE network directed to the setpoint of the fuel-flow controller. The signal passes through the Ethernet switch and on to the Linking Device for fuel flow, where it is again bridged to the Foundation Fieldbus H1 network and arrives at the fuel-flow controller setpoint. When the steam-flow PID controller is in AUTO and the fuel-flow controller is in CASCADE, these two controllers ensure that fuel flow is always correct for any steam demand. The Foundation Fieldbus network guarantees that both control loops will operate synchronously without delays. This example cannot be done with any fieldbus except Foundation Fieldbus since the synchronous timing necessary for execution of cascaded control function blocks is not supported by other fieldbuses.

In the above example, it would also be possible to use Control-Net or EtherNet/IP as the control level network to accomplish the same results. A Linking Device for ControlNet or Ether-Net/IP is available with most of the capability of the Foundation Fieldbus Linking Device. The ControlNet or EtherNet/IP Linking Device fully supports Field Control with Foundation Fieldbus H1 devices and Foundation Fieldbus function blocks.

Distributed Process Control

The Field Control example above can also be implemented as a distributed control with all of the PID control calculations assigned to a multifunction controller. In fact, many Field Control systems are first implemented as distributed control, and control loops are gradually moved to the field devices when tuning and configuration are stabilized. However, some of the popular fieldbus architectures do not support Field Control and require control only in multifunction controllers, which makes it distributed control.

Not all processes can be adequately controlled with function blocks in field devices. Some processes are too complex for all of the controls to be configured into function blocks, while others demand high-speed interaction between control functions, too fast even for Ethernet connections. Field Control is no longer new and the experience base has grown among many major petrochemical companies that have now standardized upon its use. Still, many control engineers are unwilling to risk using it. For many processes, there will be a combination of Field Control for the final control loops, typically in the control valve positioner, and all of the upper level cascade and supervisory controls in a multifunction controller.

Distillation control, as mentioned in the Distributed Process Control section of fieldbus applications, is one of those complex processes needing upper level controls in a multifunction controller. Figure 4-22 illustrates only the instrumentation at the bottom of a distillation column. A very common configuration for the bottom of a distillation column is for the bottoms level in the column to be regulated by the heat (steam) input to the reboiler to get a stable control response. The flow rate out the column bottom is then on a local flow controller. However, the most critical variables of the distillation column, such as internal reflux, cannot be directly measured but are calculated from heat and material balances. This is far too complex for field control and is a job for a multifunction controller or some other high-level algorithm in an AP (Application Processor) connected with a control level network.

Modern distillation column control in the petrochemical and petroleum refining industries uses MPC (Model Predictive Control) or MVC (MultiVariable Control) to balance the liquid and vapor flows, heat rates, and reflux ratios with column-production rates and production targets. MVC is a class of model-based controllers that take many measurement signals from the process and produce outputs to control the process. The model includes the necessary time delays to make up for long dead-time and long lag processes such as distillation. The outputs are generally not directly to control-valve positioners but to setpoints of final control loops so that the frequency of control need not be at flow-control rates. Many of the multifunction controllers of DCSs have a simple MVC in their function block library; but the capacity MVC in the DCS multifunction controllers often does not allow control of an entire distillation column. Often, the MVC is done on a separate network AP. This means that the data stream passed to and from the MVC must travel across a secure control level network such as Foundation Fieldbus HSE, ControlNet, or Profibus-DP.

Control schemes for complex processes may involve hundreds of control loops and thousands of I/O points. The time to develop these control schemes is often measured in man-years. New control systems are installed when new plants are constructed or when old control systems are modernized. The control system for the new or modernized plant will often be selected because the control scheme is already developed on a similar process, using a particular control system. The control scheme can then be transferred to a compatible new control system, perhaps with minor adjustments, but saving many man-years of development expense. There are currently no standards for the interchange of control schemes between DCSs, but DCS suppliers often provide porting tools between older and newer control systems of their own manufacture. There are no control scheme porting tools to Field Control systems. When this is the case, very often development of a Field Control scheme may cost too much.

4.3.4 Building Automation Applications

The control level network for building automation is BACnet, a high-level standard[1] designed explicitly to integrate the common networks used in modern buildings for access control, elevators, energy management, fire/life/safety, HVAC, lighting, metering, and security applications. BACnet is much like Foundation Fieldbus in its definition of 23 basic object types for intelligent field devices. While BACnet does define its own data communications protocol for the control level network, it explicitly allows WorldFIP and Profibus protocols to be used

---

1.   ISO 16484-5 (2003) Building automation and control systems --
     Part 5: Data communication protocol

since they have the capability to communicate between the host and intelligent BACnet objects.

BACnet/IP (BACnet over Ethernet TCP/IP) is the newest of these protocols and of the most interest as a control level network. BACnet/IP provides the same advantages of using commercial Ethernet as Foundation Fieldbus HSE and EtherNet/IP but for building-automation purposes. The BACnet user layer allows access directly to network objects attached to the fieldbus layer: usually LonWorks or Profibus.

LonWorks also offers its own control level network when access to LonTalk objects on the LonWorks fieldbus is necessary. LonWorks objects are defined by their profiles by the LonMark Association and are very similar to those of BACnet, but are not rationalized with them.

Review Questions, Control Level Networks, Part 2

1. Which control level networks are suitable for factory automation?

2. What is the primary application for control level networks in factory automation?

3. Which control level networks are suitable for process control?

4. What are the benefits of using a control level network to supplement a fieldbus?

5. What distinguishes a control level network from an information technology network?

Unit 5: Network Technology

This section of the book is a primer on the network technologies described in Chapter 4. Many books have been written on the details of network protocols, but they are too detailed for the task of network selection. There are some properties of network protocol that are important for network selection, and those will be covered here.

Most networks are adequate for industrial automation tasks, some better than others, but the deciding factor is usually not the protocol. The deciding factor is generally the user layer riding on top of the network, although not all network sponsors call it a "user layer."

Many networks contain features that are only useful to the systems programmer and perform some function not available otherwise. When this result is important, this chapter will call out that feature and function so that this information can be used for network selection.

There are two possible scenarios in network selection for any given project:

1. Use the network supported by your favored control system supplier, or

2. Pick the control system supplier that supports your chosen network.

This chapter will attempt to identify the major suppliers with the networks they support. In addition, some suppliers support other networks through equipment supplied by third parties. Sometimes, the third-party equipment is brand-labeled by the supplier, sometimes it's listed in a catalog of compatible suppliers, and sometimes the equipment has passed a testing laboratory's compatibility test program. This chapter will NOT identify third-party suppliers.

If you choose to use a network for applications beyond the original network design, it probably will work, but vendor support is doubtful. For this reason, many projects will have dual networks – one for process control and one for discrete I/O.

5.1. Introduction to Industrial Network Technology

All industrial networks are subject to environmental factors not included in the design of office networks. Not all networks are subject to all of the environmental factors enumerated in Table 5-1.

Table 5-1. Environmental Problems with Industrial Networks

Environmental Factor	Effect	Solutions
Heat	Expansion/contraction of connectors	Pin & socket connectors Better connector materials
Vibration	Intermittent connections	Pin & socket connectors
Corrosion	Oxides at contact points	Sealed connectors (IP67)
Electrical noise	Common mode induced false signals	High common mode noise rejection
Moving magnetic fields	Induced current	Shielding, isolation, fiber optics, wireless

Often the solution is worse than the problem. Care must be used to not demand expensive solutions for trivial problems. For example, military-style connectors solve most environmental problems with networks but at a very high cost. The most important factor is that the solution for these environmental problems be standardized; otherwise, you will not be able to "plug and play." For example, it is well known that the typical

Ethernet RJ45 connector does not perform well with continuous vibration. One of the available solutions is to use the circular M12 pin and socket connectors that have been standardized as part of the movement to industrial Ethernet. Another problem is from dust and moisture. Connectors rated to meet IEC IP67 can solve some of these problems; however, all of these connectors are much more expensive than the commercial off-the-shelf connectors commonly used for office applications.

In the early days of industrial networking, DCS and PLC suppliers offered highly proprietary networks designed to avoid the problems of mixing competitors' equipment with their own. Users soon learned that they had to become system integrators to put systems together with equipment that was not all supplied by the same supplier. The result was a software and networking nightmare. Gradually industrial networks have evolved to use more commercial hardware in areas safe to do so, and to adapt hardware when necessary at small cost increments. Software interfaces are being addressed by international standards, the OPC Foundation, consortia such as the FDT Group working on ISA103, and most of the consortia supporting the industrial network standards. There is a trend to address the software issues even more comprehensively, as the consortia for HART, Foundation Fieldbus, and Profibus-PA have cooperated and written IEC standard 61804 for EDDL (Electronic Device Description Language) independent of the communication protocol that will be used for both device configuration (an FDT function) and real-time control. This standardization work is now complete, and is also adopted as ANSI/ISA104.

As the interfaces to automation networks becomes more standardized, selection of the right network for a given project should become easier. Selection of the "right" network will then be based upon selecting the network offering the best performance and cost for a given project. However, that selection will still be strongly influenced by the networks best supported by the control system supplier. Standardization efforts should make it easier to integrate a control system with equipment supplied by the "best-in-class" supplier for each component.

Review Questions, Network Technology

1. What is the best way to choose an automation network?

2. What are the typical environmental problems that influence selecting a network?

3. What role can standards play in making network selection easier?

5.2   AS-interface

Actuator-Sensor interface (AS-i) is a bit-level sensor network that first entered the market in late 1994. For key proponent Siemens AG, AS-i completed their suite of field networks with a low-cost sensor network to complement Profibus-DP in the automation hierarchy. While Siemens was a driving force behind the network's development and AS-i is a key part of the company's field network strategy, the independent AS-inter-face Association is charged with managing the technology.

AS-i's advantage is its ability to provide low-cost electrome-chanical-connection systems for rapid transfer of messages on a 2-wire cable. By focusing on the cost and ease of installation, AS-i can differentiate itself from fieldbus competitors trying to infringe upon the sensor network market. AS-i perceives its only true competitor on the bit level to be Seriplex, due to the similarities in cost and functionality. AS-i promotes its alliance with Siemens and its position as the bit-level solution to the Siemens device network offering, but AS-i also maintains gate-ways to several other networks. Figure 5-1 shows an AS-i net-work as a stand-alone network with its own interface to a control system, and also shows an AS-i network as a subnet to a Profibus network.

AS-i is a sensor network with a message size of 4 bits that is primarily targeted at reducing the cost and complexity of wir-ing binary sensors and actuators. AS-i is positioned as a shared digital cable replacement for traditional discrete wiring of sen-sors and actuators. For industrial automation applications,

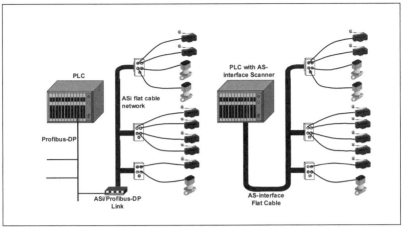

Figure 5-1. AS-i as a Subnet (Left) or as a Network (Right)

AS-i is designed to operate either as a stand-alone network or as a subnetwork to fieldbuses such as Profibus-DP or Interbus. AS-i interfaces to the Profibus-DP network through either distributed modules, such as the DP/AS-i Link, or as a direct network connected to a PLC, which in turn is connected to Profibus. Interbus is another fieldbus offering an AS-i gateway to make it into a subnet for cost-reduction purposes.

5.2.1 Technical Overview

The Actuator-Sensor interface (AS-i) is a digital, serial, bidirectional communications protocol and bus system which interconnects simple binary on-off devices, such as actuators, sensors, and discrete devices in the field. AS-i is defined by IEC in the 62026-2 standard. The 2-conductor AS-i bus cable supplies both power and data for the field devices. The AS-i bus is designed to operate over distances of up to 100 m (more if extenders or repeaters are used). No terminators are needed anywhere on the AS-i bus. The AS-i bus requires use of a special AS-i power supply that provides electrical isolation from the data signals. A special AS-i flat yellow bus cable that provides a simple cabling and connection method to most AS-i devices can be used. This cable, illustrated in Figure 5-2, is shaped so that foolproof connections with correct polarity can

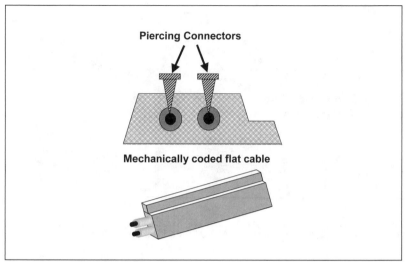

Figure 5-2. AS-i Flat Cable

be made via insulation displacement connection (IDC) technology that is built into some AS-i devices. This cabling method ensures fast connection and disconnection. Conventional round-profile cable can also be used with AS-i devices, when vendors supply screw terminal options. Many low-powered devices are powered from the AS-i yellow cable and do not require external power.

Each AS-i network can include up to 31 slave devices. Each slave can connect up to 4 conventional inputs and 4 outputs, meaning that up to 124 inputs and 124 outputs can be connected with each AS-i network segment. Network topology can include branches and stars (using passive splitters or hubs). The only limit is that the total length of AS-i cable anywhere between extenders or repeaters is limited to 100 meters. Repeaters generally require a separate AS-i power supply on the far side of the repeater.

AS-i is a deterministic master/slave network that uses a single master. The 4-bit message size limitation can be overcome for 8- or 16-bit analog signals via profiles that require either 2 or 5 cycles, respectively. Separate network interface modules are used to link devices to the network, allowing it to accommodate standard devices. Compatible devices that employ a special adapter can also be connected directly to the AS-i cable.

AS-i network modules use a vampire clamp (which they call "insulation displacement") allowing piercing connectors in the top portion of the module to pierce the flat cable insulation. Unshielded and untwisted cable is used in order to minimize cost. A variety of modules are available, including 4 inputs, 2 inputs with 2 bits each, 4 outputs, etc., as well as a module that can accommodate 4 inputs and 4 outputs. Figure 5-3 illustrates a typical AS-i module with 4 I/O ports. Note that the AS-i flat cable is untwisted and should not be run in close proximity to power line cables. Untwisted cable can couple to power lines and pick up different potentials in each conductor, causing high normal-mode differentials not found with twisted-pair cables. Round cable with twisted-pair conductors should be used if close proximity to power lines cannot be avoided. Table 5-2 contains the technical specifications for AS-i networks.

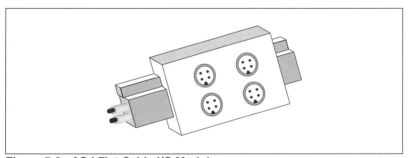

Figure 5-3. AS-i Flat Cable I/O Module

Table 5-2. Selected AS-i Specifications

Attribute	Specification
ISO layers supported	1, 2, and 7
Architecture	Master/slave with polling, single master
Wiring topology	Line, tree, star, ring
Message size	4 bits
Maximum nodes	31 slaves per master, 248 nonaddressable Inputs and Outputs
Data Transfer Rate	167 Kbytes/second
Cycle Time	5 ms

AS-i I/O modules are now available worldwide. Master units are available for most PLCs and many DCSs as well. Many types of I/O devices are available with AS-i direct interfaces, although they typically cost more than direct-wired devices. Gateways are available for the connection of AS-i networks to Profibus-DP, Profibus-FMS, Interbus, DeviceNet, CAN, Modbus, EIA-232, RS-422, and RS-485.

5.2.2 AS-i Applications

AS-i technology is being used in industrial automation and electrical-distribution applications for use with electrical switchgear and motor-control centers. Both Siemens and Schneider Electric are strong suppliers of AS-i for electrical distribution equipment and automation applications.

The majority of applications in industrial automation have been in the European automotive industry. Machine tools, welding machines, assembly equipment, and other equipment are connected with AS-i at General Motors Europe, BMW, and Daimler-Benz factories and assembly plants. AS-i is also being widely used for packaging machinery, plastics molding machinery, and in some automated machine tools.

5.2.3 AS-i Consortium

AS-International Association, which serves as the core support organization for AS-i, was formed in 1992 from a consortium of 11 European companies that together developed the specification. Today, the association has a membership of more than 60 companies worldwide, offering over 300 products and services. The North American organization is called the AS-i Trade Organization, was formed in April 1996, and serves as the worldwide promotion and member support group. The AS-i web site is http://www.as-interface.com.

5.3   CAN

Controller Area Network (CAN) is a communications protocol developed in 1989 by Robert Bosch GmbH for use in simplifying automotive internal wiring from point-to-point to a multi-dropped bus topology. Intel was instrumental in commercializing silicon for CAN in this same period. CAN has since been standardized as ISO 11898 (1993) and is being used in a number of different industry applications. In an industrial automation setting, CAN is more of a component technology, since a number of different fieldbus protocols use CAN as their base technology. Fieldbuses such as DeviceNet, SDS, CAN Kingdom, and CANopen add application layers and often modified physical layers beyond the original CAN protocol.

CiA, or CAN in Automation (http://www.can-cia.de), developed CAL, an application layer for CAN. CAL serves as the basis for the CiA-specified CANopen protocol. CAN Kingdom was developed by Kvaser (Kinnahult, Sweden) for their own use. DeviceNet and SDS were developed by a joint project of Allen-Bradley and Honeywell Micro Switch but became separate products when the joint project failed.

The primary driving force for the use of CAN in industrial automation is the low cost of the CAN chips stemming from their use in extremely large quantities in the automotive industry. While CAN is extensively used by Daimler-Benz for several of their vehicles, other automotive manufacturers have not yet implemented a CAN replacement for discrete wiring harnesses. Most of the automotive CAN networks operate at the sensor network level with no application layer. CAN silicon has been available from Intel, Motorola, National Semiconductor, Siemens, Philips, NEC, Toshiba, and Texas Instruments.

5.3.1 Technical Overview

CAN is an event-driven, multi-peer protocol. The CAN standard ISO 11898 (1993) details parts of the physical layer and the data link protocol but does not specify cables and connectors. Table 5-3 lists some of the CAN specifications.

CAN uses the producer/consumer model of data communication. Messages are sent as broadcasts on the network whenever an event indicates it should be sent. Each node on the network determines if that message should be received or not, and then either accepts or ignores it. There is no node addressing. CAN messages have an identifier field indicating both the message content and its priority. Collisions are possible in this protocol, but resolution is deterministic, using a technique called bit-wise arbitration. If a node fails, system performance may degrade, but the failure does not bring down the entire network.

Table 5-3. Selected CAN Specifications

Attribute	Specification
ISO layers supported	Partial layer 1, full layer 2
Architecture	Multi-peer
Wiring topology	Unspecified
Maximum nodes	Typically 64, no maximum specified
Message size	Up to 8 bytes
Data transfer rate	Selectable up to 1 Mbps, varies with distance

5.3.2 Application Layers

The primary differences between the use of CAN in automotive applications and for factory automation are found in the application and user layers of protocol. Application layers developed for industrial automation typically address the need for cyclic, deterministic, and synchronous communications, which are typically required to automate data flows in discrete parts manufacturing, assembly, and testing. Table 5-4 lists some of the more significant CAN Application Layers.

Since each of these CAN "extensions" adds cabling and connector specifications as well as application layers, these CAN networks are not interoperable. Both DeviceNet (Part 3) and SDS (Part 5) are parts of IEC 62026, a standard for low-voltage switchgear that also includes AS-i (Part 2) and Seriplex (Part 6)

Table 5-4. CAN Application Layers

Layer Name	Key Proponent(s)	Application
DeviceNet	ODVA	Industrial automation
Smart Distributed System (SDS)	Honeywell Micro Switch	Industrial automation
CANopen (based on CAL)	CAN in Automation (CiA)	Industrial automation
SAE J1939	Society of Automotive Engineers (SAE)	Trucks and buses
CAN Kingdom	Kvaser	Distributed control

but does nothing to correct interoperability. Each CAN physical and application layer was designed to solve specific problems. DeviceNet and SDS will be covered separately, but a brief overview of CAN Kingdom and CANopen are presented here.

5.3.3 CAN Kingdom

The Kvaser Company (Kinnahult, Sweden) began working with the CAN protocol in 1989 after five years of experience in distributed control systems. CAN Kingdom assumes that one node, the "King," is responsible for organizing the system during startup. The CAN Kingdom protocol concentrates on generating, linking, and controlling systems and does not include profiles for devices like digital and analog I/O devices. The King operates like a master but only during the configuration of the system. The King may not be involved in runtime communication between working applications in different modules. The King can then often be removed after configuration and consistency checks have been made and each module has stored its received instructions in a nonvolatile memory.

5.3.4 CANopen

CANopen is a network based upon CAN and the CAN application layer (CAL). CANopen was first used for packaging, textile, and printing machines. CANopen is now often used in material-handling applications including conveyor belts and automated warehouses. CANopen is also used in building automation (HVAC systems) and mobile applications (fork-lifts, construction machines). CAL, used in CANopen, allows integration of small sensors and actuators with PCs at the supervisory control level into one physical real-time network without requiring gateways.

5.4   ControlNet

ControlNet was developed by Allen-Bradley in early 1995. This real-time, deterministic, peer-to-peer network links PLCs and I/O subsystems at 5 Mbps. Relative to DeviceNet, ControlNet is faster, supports a larger number of nodes, and can directly accommodate longer data values such as process variables. Another application for ControlNet is peer-to-peer communi-cations between PCs and PLCs. The ControlNet system archi-tecture operates with multiple processors and has the capability to install up to 99 addressable nodes anywhere along the trunk cable of the network.

ControlNet is positioned between EtherNet/IP and DeviceNet in the automation hierarchy, but the emphasis on highly dis-tributed I/O-level communications results in an overlap with both. ControlNet uses a network model called producer/con-sumer in which each node can be a producer (sender) of data, consumer (receiver) of data, or both. ControlNet also offers plain multicast capability. This is the ability to send the same data to all network nodes at the same time. Producer/con-sumer, or a similar service on Foundation Fieldbus, Publisher/ Subscriber, uses the multicast protocol with a data identifier field allowing nodes interested in the data to quickly identify it for local use. This is more efficient than token passing or mas-ter-slave models. Time-critical data on ControlNet is determin-istically transferred during reserved time slots, whereas non-time critical data is sent during the time available after the reserved time slots.

In October 1996, Allen-Bradley placed the specifications for ControlNet in the hands of ControlNet International, a membership-supported nonprofit organization similar to ODVA that controls DeviceNet. The ControlNet specification is available to all automation suppliers in the form of a developer's guide, including a description of the ControlNet protocol, instructions on how to develop a product, and guidelines for installing and implementing a ControlNet system. In addition to the specification, development tools such as example software, a developer's starter kit, ControlNet firmware, and ASICs (Application Specific Integrated Circuits) are also available.

General Motors North American Operations (NAO) initially standardized on ControlNet and DeviceNet for its field network implementations. This choice by GM was based upon their long history of strong support from Allen-Bradley and the opening of the ControlNet technology. They have more recently modified this decision by using EtherNet/IP to replace their previous selection of ControlNet. This allows GM the flexibility to have its suppliers begin to produce EtherNet/IP compatible products. Standardization on EtherNet/IP by a major field network user such as GM, which demands open technology, accelerated the efforts of Allen-Bradley to open the EtherNet/IP technology.

At present, the future of ControlNet appears in doubt due to the broad availability and acceptance of EtherNet/IP on all of the same products. While there is no doubt about the highly synchronous messaging of ControlNet, the lower cost (to Rockwell) of EtherNet/IP and its speed allow EtherNet/IP to effectively replace ControlNet in the Rockwell product line. The application and user layers, called CIP (Control and Information Protocol) developed for ControlNet and DeviceNet are available on EtherNet/IP as well.

5.4.1 Technical Overview

ControlNet was conceived as a communications network for industrial automation applications. The cabling uses ruggedized 75 ohm 4-shield RG6 coaxial cable with passive cable taps. BNC connectors are preferred. Bus topology may be linear, star, tree, or any mix of these. Maximum bus length with repeaters is 5 Km, 1 Km without repeaters. Data rate is 5 Mbps. There may be a maximum of 99 addressable nodes or 48 per segment without a repeater. ControlNet can actually be used with any 75-ohm coaxial cable, including that used for CATV with F-connectors, but use of RG6 cable and BNC connectors provides the best rejection of industrial and radio frequency noise.

ControlNet divides time into time-critical and non-time-critical segments to allocate time slots for those data transfers that are time-critical. Each based time period (typically 20 ms) is allocated a fixed time slot (typically 12 ms) in which all time-critical producer/consumer messages are sent. The other time slot (typically 8 ms) can then be used for all non-time-critical messages. To accomplish this, the network has a coherent time synchronization. The time-slot approach, used with producer/consumer messaging, makes ControlNet highly deterministic.

The application layer for ControlNet is called CIP (Control and Information Protocol), which is the same as with DeviceNet and EtherNet/IP. See the description for CIP in the following section on EtherNet/IP.

5.4.2 Application

ControlNet is intended to be used as a control level industrial communications network connecting controllers to each other, but it has also been used to connect remote I/O units to controllers. Rockwell also markets ControlNet gateways to fieldbuses such as DeviceNet and Foundation Fieldbus H1.

5.4.3 ControlNet International

With the release of EtherNet/IP, for which CIP is available, and its assignment to ODVA (Open DeviceNet Vendors Association) for administration, ControlNet International has been given the role of guardian of the specification and little else. There are still members of the Association that build products for ControlNet, some of which are jointly marketed through Rockwell Automation's solution providers and the Rockwell Encompass program. Many of these same companies have implemented EtherNet/IP for new products but many will continue support for ControlNet for many years. Honeywell used ControlNet in their now discontinued PlantScape system, and Rockwell uses it in their ProcessLogix system.

5.5   DeviceNet

DeviceNet is one of the specific implementations of the CAN protocol in which a more complete physical layer and application layer have been specified. The DeviceNet network is a fieldbus network that provides connections between simple industrial devices, such as sensors and actuators, and higher level devices, such as PLCs and PCs. DeviceNet offers master/slave and peer-to-peer capabilities in a flexible, open network with devices from a number of vendors. Since its 1994 launch by Allen-Bradley, the DeviceNet technology has been turned over to the Open DeviceNet Vendors Association (ODVA).

DeviceNet is promoted to be a fieldbus and not a sensor network limited to connecting discrete sensors and actuators. Variable-speed drives, bar-code readers, and other devices are all used with the DeviceNet network. However, DeviceNet is usually promoted as a low-cost communications link to connect industrial I/O devices to a controller and eliminate expensive direct-wired I/O. DeviceNet is marketed as an open network since vendors are not required to purchase hardware, software, or licensing rights to connect devices to the system.

There are many DeviceNet-compatible products on the market, with Allen-Bradley the largest supplier. DeviceNet-compatible products include scanners for PLCs and single board computers, VME (VersaModule Eurocard) interface cards, PC card adaptors, pushbutton panels, variable-speed drives, remote I/O, I/O termination blocks, PCs, limit switches, proximity switches, and photoelectric sensors. Flex I/O and DeviceLink are the target Allen-Bradley interfaces for attaching standard, non-DeviceNet compatible devices. Analog signals can be interfaced through Flex I/O.

5.5.1 Technical Overview

DeviceNet is based on the ISO 11898 CAN standard, adding missing elements, such as the transmission media and application layer, which tailor the network for use in industrial automation applications. DeviceNet uses the CAN producer/consumer model for peer-to-peer communications, as well as master/slave communications where the master can either poll or strobe. Table 5-5 lists selected DeviceNet specifications.

Data-transfer rates are selectable with distance, but the emphasis has been on data rates of 500 kbps and less. DeviceNet uses a trunk line/drop line wiring topology. There are both round cables and flat cables specified for DeviceNet. Both can deliver significant DC power to the devices, eliminating the necessity for separate power wiring to remote devices, sensors, and actuators.

Table 5-5. Selected DeviceNet Specifications

Attribute	Specification
Base technology	ISO 11898 (CAN); IEC 62026-3
Architecture	Producer/consumer, master/slave
Wiring topology	Trunk line/drop line
Maximum nodes	64
Message size	Variable, 1 bit to multiple bytes
Data transfer rates by bus length	125 kbps @ 500 meters, 250 kbps @ 250 meters, 500 kbps @ 100 meters

The application layer for DeviceNet is called CIP that is the same as with ControlNet and EtherNet/IP. See the description for CIP in the following section on EtherNet/IP.

Producer/Consumer Protocol

When data is to be passed between a source such as a remote I/O device and a PLC or between a field instrument and a process controller, there are several ways in which this may occur. Polling by the PLC and controller is the most common, and is called master/slave protocol; the PLC and controller ask for and receive the most recent data on demand. Producer/consumer is a different protocol in which every field unit simply sends its data whenever it is ready. It does this by sending the data with an identifier field to allow those stations that have an interest to receive it, and the rest to ignore it. While collisions are possible, DeviceNet has such a short message length that collisions are improbable. If collisions occur, they are resolved with a deterministic method called bitwise arbitration. Producer/Consumer protocol is very low overhead, and avoids repeat transmissions of the same data to multiple network destinations.

5.5.2 Open DeviceNet Vendors Association (ODVA)

The Open DeviceNet Vendors Association, an independent, nonprofit organization, was formed in 1995 to manage the DeviceNet technology. ODVA provides developer assistance, compliance testing, promotional activities, and the maintenance and distribution of network specifications and product catalogs. Similar to other device network associations, ODVA has several special-interest groups who continue the development of DeviceNet and the EDS (Electronic Data Sheets), which are device profiles usable for DeviceNet, ControlNet, and EtherNet/IP. Anyone may obtain the DeviceNet Specification from the ODVA, and any manufacturer of DeviceNet products may join ODVA. The web site for ODVA is http://www.odva.org.

5.6   Ethernet and TCP/IP

Ethernet is the most popular LAN technology for commercial applications. Ethernet is defined by the IEEE 802.3 standard (identical to IEC/ISO 8802-3) that defines only the physical and data link layer specifications. Many network technologies have been implemented on top of Ethernet, allowing them to co-exist on the same network but not interoperate with each other. For example, the Internet uses TCP/IP, Novell networks uses a higher level protocol called IPX, and Microsoft networks uses NetBEUI protocol, all for the same purposes.

Ethernet has been used for proprietary control level networks in industrial automation for many years as well. Control system suppliers Siemens, Foxboro/Siebe/Invensys, Rockwell, Koyo/Automation Direct, and Schneider/Square-D all have had their own unique higher level protocols using Ethernet for the physical and data link layers. Remote I/O companies OPTO22, Grayhill, Burr-Brown/Texas Instruments, SixNet, Koyo/Automation Direct, and many others have offered Ethernet as one of their network technologies for several years but always with their own proprietary upper layers.

Now, Ethernet is being used within many "open" industrial automation networks, allowing suppliers to build equipment that will interoperate. Foundation Fieldbus HSE was the first of these to be specified and is included as Type 5 of the international fieldbus standard IEC 61158. Modbus/TCP, EtherNet/IP, iDA, and PROFInet soon followed, but each is directed to different applications within the industrial automation market. Only these open networks are discussed in this section.

Ethernet was invented by Robert Metcalfe while working for Xerox at their Palo Alto Research Center. At first, commercial Ethernet was implemented using a thick RG-8/U 50-ohm 4-shield cable (called Thicknet) marked at intervals where it was possible to attach a tap. The thicknet Ethernet tap clamped around the cable and penetrated the insulation and all 4 shields. This became known as a *vampire tap* for obvious reasons. Special-purpose parallel-wired *drop cable* terminated with a DB-15 connector was used to connect the tap to the computer. This was later termed 10Base5. A 50-ohm thin RG-58 cable (called Thinnet or Cheapernet) was also developed for its lower price and more convenient installation, but with fewer drops and shorter runs, and designated as 10Base2. The DIX consortium of Digital Equipment Corporation (DEC), Intel, and Xerox promoted Ethernet and eventually submitted it to the IEEE for standardization when committee 802 was formed. IEEE 802.3 was the result but had tiny frame format changes from DIX Ethernet. Today, the term Ethernet is routinely applied to the IEEE 802.3 standard, and no new implementations actually format frames in the older DIX format.

The vision of DIX Ethernet was for the main cable to snake through all offices with a drop for each office. While use of the thick main cable allowed the cable to remain intact forever, there were some reliability problems with the vampire taps. Thinnet was less expensive and used passive T-taps, but could not supply as many taps and the coaxial cable connections proved to be less reliable than desired.

Experience with the cable installed for token ring (IEEE 802.5) using a central hub wiring plan showed the more economical maintenance and installation of this cable plant. This cable plant design was adapted to Ethernet and became known as 10BaseT. The 10BaseT standard was defined in the IEEE 802.3 standard and refers to use of EIA/TIA (Electronics Industries Association/Telecommunication Industries Association) standard 568-B.2 Category 5 UTP (Unshielded Twisted-pair) cable and the RJ-45 connector.[1]

5.6.1 Technology Overview

The protocol that allows many computers to share a single Ethernet network is specified in IEEE 802.3 and is called CSMA/CD (Carrier Sense Multiple Access/Collision Detection). Each computer on the network that needs to send a mes-

---

1. Discussion of cabling standards: http://www.tiaonline.org/ marketdev/whitepapers/category6_81502.pdf

sage to any other computer simply listens for network activity, finding none, immediately begins to transmit its message. During transmission, it listens for proper data frames based on its own ability to receive bytes of data as they are sent. If another computer has simultaneously also listened and started its own transmission, the data will collide and not form valid byte frames, causing an error called a collision. Upon detection of a collision, both computers immediately stop transmitting and wait for a random time before beginning again. Since the wait time is random, only one will start while the other will now listen and find the line busy.

Since Ethernet uses a random waiting interval and collisions can occur again, especially if the network is very busy, it is not possible to determine the maximum possible wait time. This is called nondeterminism and is one of the major objections to the use of Ethernet for any time-critical task in industrial automation. Actually, in networks loaded up to 50% of the possible load, very few messages ever collide, fewer collide twice, and rarely do they collide three times; they almost never collide four or more times. The best solution for collisions is to reduce the load on the network. Luckily, 100 Mbps Ethernet now costs no more than 10 Mbps Ethernet. Just introducing the higher speed to a 50% loaded 10 Mbps Ethernet segment reduces the load to approximately 5% and makes the probability of a collision-induced delayed transmission highly unlikely.

Another way to reduce the load on an Ethernet segment is to use a switch. An Ethernet switch is a store-and-forward electronic device that buffers messages and switches them (sends the data stream) to the targeted computer. Most switches will work in full-duplex mode, meaning that it is possible to send and receive at the same time. When switched full-duplex Ethernet is used with Category 5 cable, which has separate transmit and receive pairs, there can no longer be any collisions since each cable segment contains only one computer and the port on the switch. With no collisions to resolve and the ability to send messages without ever waiting for a free network, it becomes possible to determine the maximum possible worst-case delay for a message to be delivered. This is the definition of determinism. In this type of network, message delay is determined by the speed of the switch and is not a random function. Modern switches insert negligible delays between the receipt of a data frame and the beginning of transmission to the intended receiver.

Switched full-duplex Ethernet networks also benefit from two additional factors: the maximum message length is no longer limited to 1500 bytes, and there is no minimum message length. Both of these factors were necessary to allow for reliable collision detection, but are irrelevant when collisions are eliminated. The benefits of using full-duplex switched Ethernet networks are mostly due to the commodity nature of 10/100/1000 BaseT Ethernet, a much lower cost, and much higher speeds than for most proprietary networks. The potential negative factor is that the switch is an active computing device with a measurable failure rate. Mitigating this is the fact that network switches are built to be highly reliable devices and ruggedized forms are widely available for the industrial automation market.

Ethernet switches are rated by the number of bits per second (bps) they are able to handle. Most modern switches are based on single-chip silicon solutions and have such high bandwidth (a large bps rating) that no measurable delay will come from the buffering in the switch at Fast Ethernet speeds of 100 Mbps. Latency delays are minimized since the switch determines the output port by examining the destination address in the first few bytes of the input message and streaming the remaining data directly and immediately to that port.

The IEEE 802.1D standard calls for a spanning tree bridge as a way to allow Ethernet switches to build up their own routing tables. The ports on the switch are designated as *local* or *uplink*. As messages are received from the local ports of the switch, the *from* address is noted in the routing table. At first, the switch will not have any information on the location of a message destination, so it sends that message out on all ports other than the one from which it received the message. Only the desired destination port will accept the message, since its address matches the destination address of the message. Eventually, all of the local ports will have entries in the switch's routing tables. Any message sent to an address found to not be local is then sent only to the uplink port. A layer 2 Ethernet switch uses the address of the Ethernet network interface card, but a layer 3 switch uses the IP address of the source computer.

5.7   EtherNet/IP

EtherNet/IP was created through the combined efforts of ODVA, ControlNet International, and Rockwell Automation. Earlier efforts by Rockwell to encapsulate ControlNet protocol on Ethernet/TCP/IP were the basis for this development. EtherNet/IP (Ethernet Industrial Protocol) was designed to efficiently implement data transfer using CIP at the application layer. It operates on commercial Ethernet but spawned one of the ODVA special-interest groups to investigate alternative physical wiring and connectors more suitable to industrial automation.

Through the efforts of ODVA and other, similar organizations, a standardization committee was formed by the EIA/TIA (Electronic Industries Association/Telecommunications Industries Association) to prepare specifications for industrial Ethernet cabling and connectors. Through this effort, EIA/TIA-568-B.2 Cat. 5e standard has been established. Bulkhead connectors with an RJ-45 form factor have been specified as well as a round M-12 (12 mm) 8-pin connector. The objectives for this work are to establish standards allowing industrial cable plants to be constructed according to the environmental concerns for the industrial area.

The objectives of EtherNet/IP are to provide a full industrial-grade data communications service using as much commercial off-the-shelf Ethernet hardware and cabling as possible. The benefits are to obtain the speed of Ethernet at the lowest cost possible through broad usage in the commercial market.

5.8 Foundation Fieldbus

Foundation Fieldbus was created to respond to the need for a two-way all-digital data transmission network technology for use in process control. From the beginning it was to replace the 4- to 20-mA DC transmissions previously used for analog control instrumentation. It was also to use the same type of wire typically used for analog transmission, to supply power to field instruments, and to fully conform to intrinsic safety requirements. The ANSI/ISA50.02 (1989) standard met all of these requirements and served as the basis for Foundation Fieldbus.

Figure 5-4. Fieldbus Foundation Registration Mark

The Fieldbus Foundation was formed in 1994, at the urging of many users, from two competing organizations, WorldFIP North America and ISP (Interoperable Systems Project). The users emphasized the importance of a single fieldbus protocol and the imperative of basing it upon a recognized standard. The control-systems industry had previously been divided between the prior two bus proposals. With the energy now concentrated upon a single specification, the Fieldbus Foundation rapidly completed their implementation specification based on the ANSI/ISA50.02 documents. Almost immediately, the Fieldbus Foundation began to create a testing suite to validate that field devices conformed to their specifications. The validation program was called Foundation Fieldbus Registration. Devices passing the validation testing would then be allowed to carry the Foundation's registration symbol as illustrated in Figure 5-4.

5.8.1 Wiring and Signaling

The initial Foundation Fieldbus specification was for H1, for the targeted instrumentation connection 4-20 mA replacement application. H1 (stands for Hunk 1) operates at 31,250 Kbps, a very low speed for a communications bus, but necessary because of the need to reject noise, deliver DC power, and provide intrinsic safety. Noise rejection is enhanced through the use of a trapezoidal waveform, rather than the traditional digital square wave. Encoding is Manchester Bi-phase that requires both a high and low state for each bit. The signal is differential between the two wires rather than the more noise-prone single-ended signals of EIA-232. The wire is a shielded single twisted-pair in which the impedance is not specified to allow conventional analog instrument cable to be used.

The topology of Foundation Fieldbus H1 is called trunk and spur. In development of the 50.02 standard, it was called a *chicken foot*, because of the way it was always illustrated, as in Figure 5-5. There are very few wiring restrictions for H1, allowing the cable to be installed in the most economical way. The chicken foot illustration is most often used since it allows easy maintenance, but daisy-chain wiring from instrument to another instrument is also allowed. In any case, the wiring from the H1 junction box is usually a shielded multicore cable to continue the wiring from each instrument to the H1 termination I/O card. However, since H1 instruments are addressed by a bus address, all of the field devices of a single segment are connected together inside the junction box and use only one pair of the multicore cable. This is illustrated in Figure 5-6.

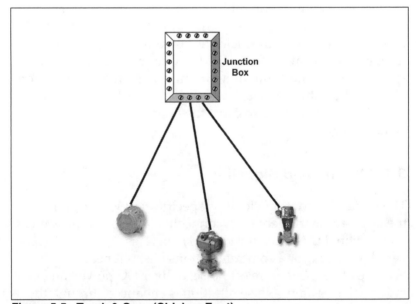

Figure 5-5. Trunk & Spur (Chicken Foot)

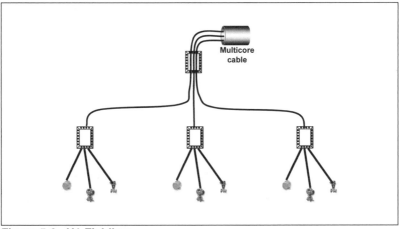

Figure 5-6. H1 Fieldbus

5.8.2 Intrinsic Safety (IS)

Intrinsic safety is supported by H1 but not required. The basic requirement for intrinsic safety is a barrier located where the wiring passes from a safe zone into the hazardous zone. The barrier ensures that no electrical current with enough energy to ignite a flammable gas mixture can cross the barrier but is instead shunted to earth ground. Since Foundation Fieldbus H1 is a differential signal, the intrinsic safety barrier is bi-polar, unlike that for an analog signal where one side is usually referenced to an earth ground. The barrier limits the voltage that can be transmitted on the H1 link and therefore also the power available to be used by the field instruments connected. The original H1 specification limits the number of units connected with a single fieldbus segment only to the number of available addresses (254) but does limit the maximum current draw for intrinsic safety. As field device power requirements are decreased with low-powered electronics, the number of units per segment can be increased to maintain intrinsic safety. The revised H1 specification is based on the European-developed FISCO (Fieldbus Intrinsically Safe Concept) specification that refers to the *actual* power consumption for the field devices actually being used. This is often significantly lower than the maximum power requirements, and allows about twice as

many field devices per intrinsically safe Foundation Fieldbus
H1 segment.

> ### Intrinsic Safety (IS)
>
> **IS is defined by ANSI/ISA-RP12.06.01-1995 as "Equip-
> ment and wiring which is incapable of releasing suffi-
> cient electrical or thermal energy under normal or
> abnormal conditions to cause ignition of a specific haz-
> ardous atmospheric mixture in its most easily ignited
> concentration." IS is achieved by limiting the power
> available to the electrical equipment in the hazardous
> area to a level below that which will ignite the gases.**

Later development of the FNICO (Fieldbus Non-Incendive
Concept) for installation in Zone 2 areas where flammable
gases are not normally found, uses identical concepts as IS, but
allows considerably more energy per fieldbus segment. FNICO
allows about three to four times the number of devices per
fieldbus segment as does IS.

While DC power for the field device can be extracted from a
Foundation Fieldbus H1 segment, it is not required. Devices
may be self-powered much as they were in analog control systems. Devices powered from the fieldbus must operate within
9-32 V DC. Most commercial fieldbus power supplies operate
at approximately 24 V DC. The maximum number of devices
that can be powered from the H1 bus depends upon the use of
intrinsic safety or not. Typically, about 6-9 devices can be powered from an intrinsically safe FISCO-conforming fieldbus.
Nominally, about 30 devices can be powered from a non-intrin-
sically safe fieldbus. However, good engineering practice limits
the number of devices per H1 fieldbus segment to less than
either of these limits.

5.8.3 Fault Tolerance and Single-Loop Integrity

Each H1 fieldbus segment should not connect elements required for more than one control loop. Typically, this means about 2–3 devices per segment plus any information-only measurements not used in control. The rationale for this is called *single-loop integrity.*

Single-loop integrity is a concept based on the original use of single-loop controllers that were originally mechanical, then pneumatic, then analog electronic, and finally digital electronic. The philosophy was that any failure would cause the loss of only a *single loop of control.* It was this philosophy that allowed these controllers to be used for the regulation of critical processes without requiring redundancy. Actually, most instrument engineers didn't even think about this aspect until the creation of the DCS and its *shared controllers.* With shared control for critical processes came the need for fault tolerance, since each controller was now responsible for many control loops. The most common technology for fault tolerance was the use of redundant controllers with no single points of failure.

Foundation Fieldbus H1 was conceived for the purpose of returning to single-loop integrity when required. To do this means to configure each H1 segment with the measurement transmitters, actuators, and controllers for only one loop of control. When this is done, there is no longer a need for fault tolerance since *by definition,* single-loop integrity does not require fault tolerance. However, many processes do have indicating measurements that are not part of control loops, and these are typically configured into the single-loop H1 control segments. For reasons of economy, H1 segments for noncritical controls are often configured with multiple control loops.

Foundation Fieldbus was created to allow even complex 2- or 3-level cascade controls to be implemented totally within the field instrumentation. Often, the final control loop is configured into the process control valve regulating a flow. Usually, the value being controlled is not just the flow but is a level or temperature in the process that has its own controller. When this is the case, it is common practice to configure the measurements for the upper level temperature or level control and for the flow control into the same fieldbus H1 segment. The temperature controller is then configured into the temperature measuring instrument and its controlled output is cascaded to the setpoint of the flow controller. The cascade signal is sent or cascaded using the Foundation Fieldbus H1 segment. This is illustrated in Figure 5-7. If there is only Foundation Fieldbus H1, then there is no other way to close this cascade except for a software support function of the DCS multifunction controller that is outside the scope of the Foundation Fieldbus specification. See the section on Foundation Fieldbus HSE for the "official" way to link H1 segments.

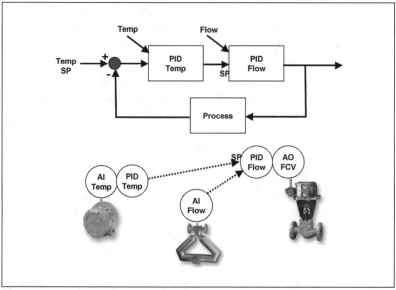

Figure 5-7. 2-Level Cascade Control

Single-loop integrity is not the only answer to fault tolerance with Foundation Fieldbus H1. While few if any Foundation Fieldbus instruments have dual ports to allow redundant H1 wiring, for super-critical control loops it is possible to duplicate every element of the control loop except for the control-valve actuator. Many DCSs also offer dual-ported H1 interface cards for this purpose where dual H1 segments are configured. There are many ways to use H1 redundancy to achieve the level of fault tolerance required for any control loop, but at considerable expense. With several years of commercial experience in using Foundation Fieldbus H1, it is now safe to conclude that this communication link is not a common source of failure. Much higher failure rates are experienced with process control instrumentation, usually associated with the sensor itself rather than the digital electronics used to transmit or receive signals and data. For example, installing two differential pressure transmitters on one set of impulse lines is not a good idea. A better idea would be to install a vortex shedding flowmeter to back up a differential pressure/orifice flowmeter.

5.8.4 Protocol

The internal communications controls of Foundation Fieldbus are all contained in the data link layer and application layers. Mostly this is invisible to the user, but it may be one of the factors used in network selection, although not as important as the wiring and signaling. The actual protocol will not be described here, only the effect of that protocol that is important to network selection.

Foundation Fieldbus H1 was designed for use in process control. Explicitly, it was designed expecting that many of the control loops needed for process plant control will be done in the instrumentation, transmitters, and actuators and used in all process control. Further, it is designed to implement time-critical cascade loop controls totally within the field devices without contribution by a host device except for initialization and supervision. This has been named *Field Control*. The most important aspect of field control is that the network itself enforces the critical timing, ensuring just-in-time delivery of measurement data for use by the feedback-loop controls. While this can be accomplished in many ways, Foundation Fieldbus uses the method developed in the ANSI/ISA50.02 standard called bus arbitration, using the Link Active Scheduler attribute of most fieldbus devices. This protocol does not require a host device or any additional equipment, yet it makes the entire network time synchronous for control purposes.

The data flows of Foundation Fieldbus are controlled by the field device, not by a host device, to ensure that network loading remains prioritized about the needs for cascade loop control. The data-flow mechanism is called Publish/Subscribe, in which all elements needing access to real-time data subscribe to that data with its source, usually the field device. When the data is ready for that subscription, it is published on the network for the subscriber. Publish/Subscribe is a refined form of the producer/consumer data flow used by DeviceNet, ControlNet, and EtherNet/IP.

5.8.5 Function Blocks

Foundation Fieldbus function blocks are derived from the work of the ISA50 Fieldbus standards committee as documented in ISA TR50.02. They are in turn taken from the typical function blocks of DCS and computer control before that. They bear no direct relationship with the function blocks described by IEC 61131-3, Function Block Diagramming. Furthermore, only portions of IEC 61499 apply to the description of Foundation Fieldbus function blocks. The differences are that 61131-3 and 61499 describe blocks of programming that are executed in sequence as part of the program of a controller. Foundation Fieldbus function blocks are static entities that execute in parallel, which is possible because they are usually in totally separate devices. There is no concept of sequence of execution, rather execution is cyclical based on precise time intervals. Think of a Foundation Fieldbus function block as an instrument part of a control loop. The only sequence is the flow of data supplied by upstream blocks to downstream blocks.

Foundation Fieldbus function blocks are modeled after function blocks in a DCS, which in turn are based on functions previously done in function blocks of computer control systems. While many function blocks are based on corresponding functions of analog controllers, most of the blocks are based on decision-making and dynamic process modeling. The Fieldbus Foundation attempted to standardize on the most used function blocks that are listed in Table 5-6. The performance of the first 10 of these blocks is tested by the Foundation when

Table 5-6. Foundation Fieldbus Standard and Advanced Function Blocks

Function Block Name	Symbol
Analog Input	AI
Analog Output	AO
Bias/Gain	BG
Control Selector	CS
Discrete Input	DI
Discrete Output	DO
Manual Loader	ML
Proportional/Derivative	PD
Proportional/Integral/Derivative	PID
Ratio	RA
Device Control	DC
Output Splitter	OS
Signal Characterizer	SC
Lead-Lag	LL
Deadtime	DT
Integrator (Totalizer)	IT
Setpoint Ramp Generator	SPG
Input Selector	IS
Arithmetic	AR
Timer	TMR
Analog Alarm	AAL
Multiple Analog Input	MAI
Multiple Analog Output	MAO
Multiple Discrete Input	MDI
Multiple Discrete Output	MDO
Discrete Alarm	DAL
Flexible Function Block	FFB
Calculate	CAL
Analog Human Interface	AHI
Discrete Human Interface	DHI

devices are registered and are thus uniform in industry. However, manufacturers always seek to make their devices with competitive advantage and add new function blocks of their own design. Foundation specifications allow this, provided that the new function block is supplied with an Electronic

Device Description (EDD) that enables any other block or host application to communicate with the function block through its attributes.

Electronic Device Descriptions are written in EDDL (EDD Language), which are text files describing in detail the data structures of each of the function blocks' attributes. EDDs also exist for all of the standard Foundation Fieldbus function blocks. Foundation Fieldbus EDDs are very similar to HART DDs and are functionally similar to GSDs defined for Profibus. The three organizations supporting these protocols have "harmonized" the formats for the specification of their device descriptions in areas where they overlap and issued this as IEC 61804 Function Blocks for Process Control, which is also ANSI/ISA104.

Only Foundation Fieldbus includes function blocks for feedback loop control consistent with process control applications. Profibus-PA includes "profiles" for I/O properties that are similar to the AI and AO blocks in Foundation Fieldbus. However, Profibus-PA was not designed for direct communications between field devices, necessary for cascade formation or for closed loop field control.

5.9   Foundation Fieldbus HSE

HSE is the backbone or control level bus for process control using Foundation Fieldbus H1 for transmitters and actuators. Originally, as defined in IEC 61158 Part 2 and the ANSI/ISA50.02-2 standard, there was an H2 (Hunk 2) for the higher level bus functions. However, the AC bus at 1 Mbps, the voltage mode bus at 1 Mbps, and the fiber-optic buses at 1 and 2.5 Mbps were all found to be too expensive and too slow. Instead, Foundation Fieldbus HSE was created to use commercial off-the-shelf (COTS) Fast Ethernet and Internet software standard protocols. This created a very fast bus (100 Mbps) as well as providing the economies of scale from Ethernet. Foundation Fieldbus HSE is Type 5 of the IEC 61158 fieldbus standard.

Process control networks based on use of only Foundation Fieldbus H1 must have each bus segment terminate at a process controller or some other similar device. This does not allow a device on one segment to communicate directly with a device on another segment. Many process controllers offer a communications service to allow these communications, but they cannot be guaranteed to perform the data transfers fast enough for use in cascade loop control. The plan for Foundation Fieldbus was always to have an H2 level bus to join segments directly; HSE is that bus. Part of the Foundation Fieldbus architecture includes Linking Devices to join several H1 bus segments to one or more HSE networks. This architecture is illustrated in Figure 5-8.

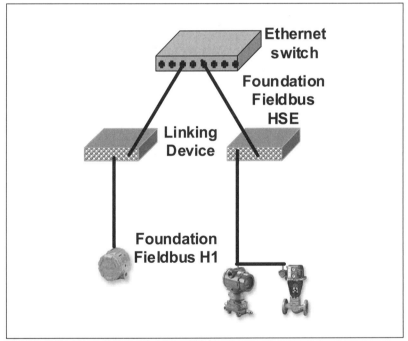

Figure 5-8. Foundation Fieldbus Wiring

5.9.1 Technology Overview

Foundation Fieldbus HSE uses the same application and user layers as H1 and so completely interoperates with it. The Linking Device performs the spanning tree bridge functions specified in the data link layer of the ISA and IEC fieldbus standards and so fulfills all of the requirements for the H2 bus. However, instead of mapping the application layer functions to the H1 data link layer, HSE maps the fieldbus functions to UDP/IP (User Datagram Protocol/Internet Protocol) data frames carried on standard 100BaseT Ethernet. This means that you use commercial everyday Ethernet wiring, accessories, and terminations where environmental conditions permit. In other locations, industrial Ethernet wiring and components should be used.

HSE is designed to provide fault tolerance using cable redundancy. Since HSE is to be used as part of the closed-loop control network, real-time message delivery is critical. Where the control data message passes over HSE, that cable segment should be redundant. The HSE redundancy scheme does not wait for cable failure and switchover, since that cannot guarantee real-time message delivery. HSE redundancy requires that the same message be sent on all bus segments at the same time with the same message identity. Only one such message is used; the other is intercepted with a redundancy manager. Failure to receive a redundant message on a redundant segment indicates a cable failure. More than dual redundancy is supported.

HSE supports exactly the same software interface as Foundation Fieldbus H1 and therefore has all of the same features as H1, except for intrinsic safety and delivery of power to field equipment. With the mid-2008 announcement of the first intrinsically safe Ethernet switch and a Wi-Fi access point[1], both capable of supplying power over the Ethernet wiring using the IEEE802.11af standard, this situation is likely to change. Field device manufacturers may now build equipment communicating on HSE and drawing power from the Ethernet wiring, while remaining intrinsically safe.

---

1. MTL Instruments division of Cooper Crouse-Hinds, Series 9400. http://www.mtl-inst.com/newsroom/press_releases/pr474.htm

5.10 HART

HART stands for Highway Addressable Remote Transducer. It was created to allow digital field instruments to communicate data to host systems, while simultaneously transmitting ANSI/ISA50.1 standard 4-20 mA. Since a digital processor is required in the transmitter, many parts of the measurement are now processed in software replacing more expensive hardware and analog circuitry. This has lowered the selling price of HART transmitters to parity with pure analog electronics. Customers with plans for the future tend to use HART transmitters on all new processes and in retrofits of existing plants.

HART enables range changing. Modern instruments are designed to operate over much larger ranges than older analog instruments that tended to build the range into the design in order to achieve high accuracy. Digital transmitters are able to use a portion of the range of a wide-range sensor and still maintain high accuracy. This allows the range to be changed through software as long as the narrower range is within the limits of the wide range. Furthermore, any nonlinearity due to the wide range can be compensated in software. Figure 5-9 illustrates the range change that might be used for a HART differential pressure transmitter.

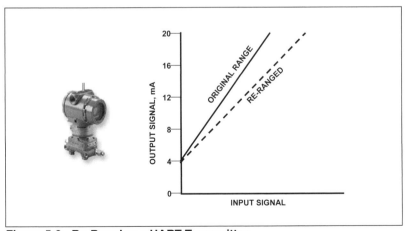

Figure 5-9. Re-Ranging a HART Transmitter

Dynamic range changing has broad economic effects. Users and distributors can stock only one spare unit to serve in many applications, changing the range before use. Changing the range and setting the zero point during operating conditions can be done from ground level, rather than sending a maintenance tech to the top of a 100-m tall distillation column. Range changes are logged at the instrument itself, along with the identity of the maintenance tech, and the date. The instrument serial number, manufacturer, model number, etc. are all readable using the HART protocol.

A HART transmitter looks very much like an analog field transmitter. The wiring is the same, and the measured value between 4 and 20 mA is the same once the range has been set. Similarly, HART control-valve positioners look identical with analog valve positioners and accept the same 4-20 mA signal. However, where analog transmitters and control-valve positioners require screwdriver adjustment to zero and span, HART devices are configured with a HART handheld communicator, through the DCS or with a PC-based device.

HART devices usually have secondary variables available, such as the instrument case temperature, which are useful for maintenance purposes. The instrument may also have significant variables such as upstream line pressure on a differential pressure transmitter, or valve stem feedback position for a control-valve positioner. These data are accessible only through the HART digital protocol.

One of the recognized but not too widely used features of HART devices is their ability to record the original instrument configuration, including a bill of materials and succeeding changes such as calibrations, repairs, and service dates. This information is typically entered using a HART handheld terminal by the maintenance or installation technician. It is readable through the HART I/O interface card for the DCS. This is the type of information required for asset management and computerized maintenance management software.

5.10.1 Technology Overview

HART is implemented by mixing a low level digital current loop signal with the 4-20 mA DC primary value signal. The mixing or modulation uses FSK (frequency shift keying) as in the Bell 202 modem specification, which limits the speed of the digital transmission to 1200 baud. This speed and modulation allows HART to use conventional intrinsic safety barriers.

Often, HART devices are used with conventional analog I/O equipment using only the single 4-20 mA primary value signal, and the digital content is ignored. In this case, a handheld terminal is used to set up and maintain the device. The handheld unit is usually attached temporarily to the termination connections at the DCS or other controller, or it may be attached at any other point in the wiring between the controller and the field device. Attachment of the handheld at the field instrument is often required during zero setting or range calibration.

The most popular HART handheld unit is the Model 375 sold by Emerson Process Management, and resold by most HART instrument suppliers. The HART instrument data structure is defined by a device definition (DD) that is supplied by the manufacturer of the instrument to the HART Communications Foundation that maintains a library of registered DDs. This library must be loaded to the 375 handheld so that the parameters may be displayed and adjusted. Similar handheld devices are available from other suppliers. A list of all HART instruments registered with the HART Communications Foundation may be seen on the web at http://www.hartcomm.org — look for the Product Catalog.

Modern DCSs offer direct HART I/O interfaces, meaning that the DCS can read and adjust all of the HART instrument parameters and play an active role in zeroing and calibration. The analog value is usually used for control purposes since the interface has analog input capability, but any secondary variables are read digitally through the HART digital interface. A single HART device can have as many as 256 different values stored in binary form including floating-point variables. The

HART DD library must be downloaded to the DCS to allow it to access HART devices.

A multidropped version of HART also exists as a purely digital transmission method. In this case the analog primary variable signal is suppressed. Up to fifteen HART field devices can be connected to a single HART digital loop. The speed remains at 1200 baud. Although the HART specification provides for a higher speed HART digital loop running at 9600 baud, this higher speed digital loop is not often used. The controller polls each HART device for data. Multidropped HART is not often used since the data rate is too slow.

5.11 iDA

Interface for Distributed Automation (iDA) is a protocol originally backed by Schneider Electric, Jetter, Phoenix, Wago, Sick, Lenze, Turck, Innotec, and RTI. The objective of iDA was to use Ethernet as a time-critical network for motion control and other discrete manufacturing applications. These supporting companies view iDA as the next evolutionary step for Modbus/TCP and Interbus. The protocol is completely open and published on the web at:

http://www2.modbus-ida.org/idagroup/service/download/
IDA-Spec-V11.pdf

Commercially, iDA is not yet available, but its features are expected to eventually become available in products from its supporters. As the strategic successor to the very popular Modbus/TCP, iDA is expected to eventually attain broad market acceptance.

The organization of iDA strongly supports object-oriented function block programming defined by IEC 61131-3 Programmable controllers - Part 3: Programming languages. This standard uses function blocks in several of its five sections. It should also be noted that Schneider Electric favors the use of Sequential Function Chart programming for all of its PLCs and provides function block elements in its Ladder Diagram programming as well. Therefore, iDA is viewed as the protocol to allow networks of PLCs to interoperate.

Since the application reference for iDA has been for motion control, the most demanding function in discrete manufacturing, it should prove more than adequate for all applications in manufacturing. Motion control requires high speed and highly synchronous data transfers when used for multiple axes of control to ensure exact placement of parts and precision machining.

5.11.1 iDA Safety Bus

Many of the automation bus technologies have added capabilities to be used in safety systems. iDA was originally designed to include safety applications. The basic principle for safety on iDA is to embed the safety decision in the final actuator. The primary mechanism used for safety in iDA is to embed a safety frame within the iDA transmission. The safety frame contains redundant and inverted data sent from the interlock source to the safety actuator and additional error checking to ensure delivery of error-free data beyond the normal error checking of Ethernet. Additionally, the safety data frame includes a sequence number proving the source node to be alive and operational. Since all safety data is sent as time-critical data, an RTPS (Real-Time Publish-Subscribe) mechanism is used that provides data synchronization that can be tested by the safety actuator. Finally, a fail-safe watchdog timer is used in case all else fails. The watchdog timer is maintained at the safety actuator and is tripped if safety data is not updated before the timer expires.

5.11.2 Technical Overview

iDA is defined as a RTPS protocol using the specification developed by Real Time Innovations, Inc. The specification for RTPS has been submitted to the IETF (Internet Engineering Task Force) but there was insufficient interest to publish it as an application standard. The complete description of RTPS proto-

col is available on the web at http://www.rti.com/docs/
RTC_Feb02.pdf.

Publish/Subscribe is a refined form of producer/consumer
data distribution for networks. In producer/consumer systems, new data is multicast with an identifier whenever it is
changed, or on a planned schedule. Publish/Subscribe requires
that at least one "subscriber" has registered with the source of
data and declared the frequency at which it requires an update.
Data is then multicast at each subscribed interval with an identifier enabling those network nodes needing the data to quickly
accept it. RTPS adds the self-clocking and network-clock synchronization to every publishing node. Clock synchronization
allows the publishing of data to be used as a critical timing
event for synchronous applications such as PID calculation for
motion control or process control.

iDA defines communications in terms of function block objects,
and so strongly integrates with standard IEC 61131-3 Function
Block Diagramming and Sequential Function Chart
programming. The iDA function block model is based on the
standard IEC 61499 Function Blocks for industrial-process
measurement and control systems. Real-time data access is
organized by establishing publishing and subscribing blocks.
The RTPS mechanism then carries out the specifications,
transferring data on schedule using highly efficient UDP over
Ethernet protocol and IEEE 802.1d and 802.1p standards for
switching and multicasting.

All of the features of Modbus/TCP are tightly organized into
iDA to maintain both a data on demand capability as well as
backwards compatibility. Figure 5-10 from the iDA specification illustrates the organization of the protocol.

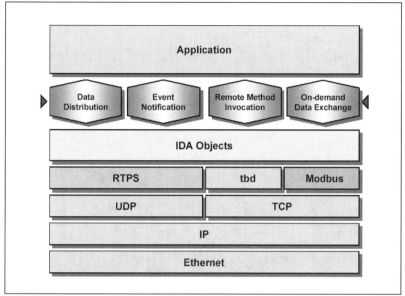

Figure 5-10. iDA Real-time Communication Services

5.12 Interbus

Interbus is designed to be a highly efficient fieldbus and an integrated sensor network called the local loop. Figure 5-11 illustrates Interbus topology. Field devices (sensors and actuators) are typically connected to a local loop I/O module. The I/O modules are connected to each other in a loop, receiving data from previous I/O modules and sending data to the next I/O module in the local loop until the loop ends in a loop termination module.

The Interbus fieldbus interconnects all remote bus modules including the local loop termination modules together into a ring. The last I/O module in the Interbus remote bus closes the ring, returning the signal to the master.

The maximum I/O count is 4096 points per Interbus network made up of both local loops and I/O terminated on remote nodes. There is a maximum of 512 local and remote modules, of which there can be a maximum of 192 local loop modules. The data rate is 500 kbps, and the maximum bus length

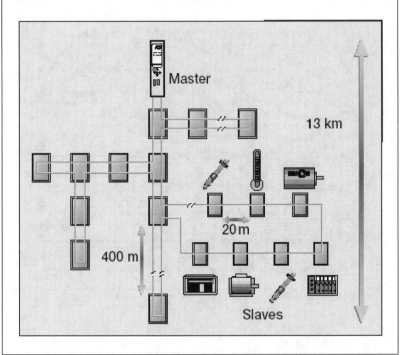

Figure 5-11. Interbus Topology (Source: Interbus Club)

between any two remote bus modules is 400 m. Since each remote bus module includes its own repeater, very long networks, up to 13 km, can be configured. Local loop modules can be a maximum of 20 m apart.

Interbus modules are available for a variety of functions such as variable-speed drives, motor contactors, motion controllers, encoders, and barcode readers, as well as for analog and digital discrete I/O. The Interbus Club web site lists products implementing Interbus at http://www.interbusclub.com.

Phoenix Contact created Interbus in 1988 to reduce the cost of wiring for discrete automation. In 2000, Interbus became Type 8 of the IEC 61158 Fieldbus standard. In 2002, Phoenix Contact became a working partner in the production of the iDA Group to ensure that iDA protocol could become a successor to Interbus for high efficiency applications such as motion control.

While the millions of Interbus nodes will be supported for many years, it appears that the future of Interbus is tied strongly to iDA. Expect that iDA nodes for Interbus loop will become available to reduce the cost of iDA for discrete I/O.

5.12.1 Interbus Technical Overview

Interbus gets its high efficiency by making all data appear to be a single shift register with the I/O data assigned to data slots associated with the ring wiring order. The master begins each cycle with output data intended for actuator modules located in the actuator slots, and the previous input data located in the slots for input modules. The long data frame is passed from each module to the next around the ring with each output module using the data located in its slot and input modules updating the data in their own slot. When the last module in the ring is reached, the entire data frame is returned to the master through each remote module, which acts as a simple repeater. Using a slotted ring shift register data structure achieves very high efficiency since there is very little protocol overhead. Cycle times of 2 ms are common for networks with maximum I/O count.

Network nodes with longer data sets, such as analog input or output and other devices needing parametric data, can also be slotted into the ring data structure using data segmentation. The longer data sets are partitioned into shorter segments and slotted into the ring structure one segment at a time. The results are reassembled at the master or remote destination node using several data frames to transport the data.

All communications for Interbus are handled by the same protocol chip, wired differently for remote nodes or local loops. All nodes receive the ring shift register and process the local data and pass it on to the next node. The return channel is used for the remote nodes but not on the local loop.

Both the remote bus and the local loop typically use RS-485 signaling over Cat-5 twisted-pair copper wiring to provide an extra pair to supply power to the modules. A single twisted pair in each direction is used for the remote bus (forward and return paths or two-pairs) and only a single twisted pair (forward path only) is used in the local loop. Fiber optics can also be used to increase the distance between remote nodes when necessary.

5.13 LonWorks

The LonWorks system was originally developed by Echelon Corp. in the late 1980s to be a low-cost and moderate-performance network for residential, building, commercial, and industrial automation. It has succeeded in all of these markets and dominates the building automation market. While originally developed for a simple 2-wire twisted-pair network, alternate media such as power line modulation, fiber optics, radio, and infrared have always been offered. LonWorks power line modulation is probably the most popular alternative media in actual use. Great progress in wireless LonWorks has also been demonstrated with constantly reduced cost.

The protocol for LonWorks is called LonTalk and was originally held as a trade secret of Echelon, but it now has been standardized as ANSI/EIA 709.1. The entire protocol, all seven ISO layers, is implemented in silicon on *neuron* chips functionally designed by Echelon but produced and sold by Toshiba and Cypress Semiconductor. Each neuron chip has three microprocessors to handle the protocol, the media modulation, and the application. Simple applications such as I/O processing can be accomplished using only the microprocessor power of the neuron chip. LonTalk is also the basis for IEEE 1473-L, a standard for rail transportation communications.

LonWorks is a peer-to-peer network intended for linking clusters of I/O to a controller. While this objective is similar to CAN, the applications for automation are more demanding and therefore require much greater microprocessor capacity on the neuron chip. The cost of the neuron chip is about triple the cost of a CAN chip, reflecting the greater capability, but the neuron chip often eliminates the need for a local microprocessor at the node.

With the opening of the LonTalk protocol, it becomes possible to port the protocol to chips other than the neuron. While the potential exists, there are currently no other implementations of chips supporting the ANSI/EIA 709.1 protocol.

Interoperability of LonWorks devices is the responsibility of the LonMark Interoperability Association, which offers a battery of tests for interoperability. The LonMark Interoperability Association web site (http://www.LonMark.com) lists thousands of products that are certified to carry the LonMark logo, the symbol of LonWorks interoperability.

LonWorks networks can be connected to Internet and other TCP/IP networks by means of the i.LON™ 1000 Internet Server, a Cisco product. The i.LON is a LonTalk/IP router enabling devices on a LAN to communicate directly with devices on a LonWorks network.

5.13.1 LonTalk Technical Overview

ANSI/EIA 709.1 is the defining protocol document for LonTalk. The problem in all multipeer networks is to control network access to prevent two or more stations from talking at the same time. LonTalk is a collision avoidance CSMA (Carrier Sense, Multiple Access) protocol using a predictive p-persistent media access control to prevent collisions even during periods of heavy loading. Rather than using a random backoff period like Ethernet, LonTalk randomizes the backoff in case of a collision into 1 of 16 different levels of delay. Since the maximum delay period is known, this method of collision resolution is deterministic. Further, the algorithm minimizes any access delays during periods of light loading. In addition, a master station can be used to poll all nodes for values, effectively eliminating all collisions and exactly controlling access timing.

Network traffic is minimized by allowing direct communications between network nodes rather than relaying through a

master station. In many residential, building, and even indus-
trial-control applications, direct access between nodes eliminates the need for a master controller and lowers cost of the
system.

LonWorks has been implemented over a wide variety of physical media including twisted-pair copper wiring, power line
carrier, and wireless (radio). Most common is the free-topology
link that allows any combination of bus trunk or branched tree
wiring to communicate over a maximum distance of 500 m at
speeds up to 78 kbps. To get longer distances, a bus topology
can be used up to 2200 m. To get higher speeds, a bus topology
is required for speeds up to 1.25 Mbps for lengths up to 125 m.
The maximum number of network nodes is 64, except on the
free topology system where up to 128 nodes may be configured. The power line carrier media is limited to a 5.4 kbps data
rate, but length is limited only by the conductivity through the
power line medium.

5.14 Modbus

Modbus in all of its forms is the most popular control level bus.
It was originally created by Modicon as a means for computers
to gather information and control the operation of their PLCs.
The data organization of all PLCs is as a set of addressable registers organized into sets as I/O, control relay, analog inputs,
analog outputs, and variables. PLC I/O is understood to be
organized so that each digital discrete input device appears as
a single bit in the I/O registers according to its location in the
I/O hardware. This usually means that digital discrete output
is often mixed together in some of the registers, requiring a register mask to define the outputs.

Control relays are the internal bit variables used for intermediate results of logical decisions. A control relay has by default as
many normally open or normally closed discrete results as
required, but only the control relay itself is represented as a bit
in a register.

Analog input, output, or variable is represented by a whole 16-bit register. Initially, the PLC only had A/D or D/A converter results, but increasingly microprocessors are handling these operations in floating point within function blocks, and these results are represented in a set of floating-point registers assigned by the programmer.

Modbus commands provide ways to transfer the content of one or many registers from the PLC to the host device, which may be a computer or may be another PLC. The Modbus command set was so popular when it was first created in 1979 that it has often been copied for other PLCs. The most popular Modbus spinoff is J-Bus, which has been used by Siemens, Telemechanique, and many other smaller PLC suppliers as a secondary access protocol on many PLCs not originating from Modicon roots. The idea behind Modbus, a command set operating on 16-bit registers, has been used by all PLC suppliers and by ISO 9506 Manufacturing Message Specification (MMS). Additionally, Modbus continues to be the most common protocol for use in SCADA (Supervisory Control And Data Acquisition). The same structure is used in OPC/DA.

The identical command set for Modbus has been implemented using several different physical layers. Table 5-7 shows the physical layers used for Modbus, which are also illustrated in Figure 5-12.

Table 5-7. Modbus Physical Layers

Physical Layer	Standard	Speed	Comments
Modbus	EIA-232	19.2 kbps	Original serial Modbus
	EIA-485	Up to 1 Mbps	High-speed serial Modbus
Modbus+ (Plus)	HDLC (ISO 3309) on EIA-485	Up to 1 Mbps	Token passing multipeer bus. Proprietary protocol.
Modbus/TCP	IEEE 802.3	10/100/1000 Mbps	Modbus on Ethernet

Modbus was originally developed for operation on an asynchronous serial line, now defined by standard ANSI/EIA/TIA-

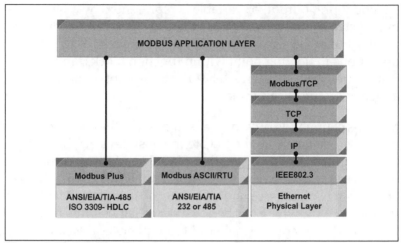

Figure 5-12. Modbus Structure

232F. This made it compatible with serial port modems without further definition. When remote termination units (RTU) were developed for SCADA systems, Modbus was a natural selection for the transport protocol across serial dial-up communications lines. This version of Modbus has become known as Modbus/RTU, and is still very popular for many applications, not just RTUs.

In local plants, where many PLCs need to be connected to a single computer, Modbus/ASCII is considered too slow, and does not support multidrop communications. For these applications, Modbus has been implemented on an ANSI/EIA/TIA-485 multidropped serial bus as Modbus/RTU. EIA-485 uses a balanced differential line enabling much longer distances, much higher speeds, and better noise rejection than EIA-232F on a twisted-pair cable. This continues to be a very popular choice for implementation of Modbus.

Schneider/Modicon also supported a version of Modbus called Modbus Plus or usually *Modbus+*. The specification for Modbus+ has never been formally released as an open network, but it is still used on Modicon-brand products as well as those external products developed with Schneider agreements. Modbus+ uses a serial line with HDLC (High-level Data Link Control) protocol that enables multidrop communications.

Modbus/TCP was developed in 1998 and was declared to be "open." The specification has been published on the Modicon/Schneider web site, but ownership has now been transferred to the independent Modbus.org association. The specification is published on the association's web site: http://www.modbus.org. Modbus/TCP enables several improvements to Modbus communications: It lowers cost by using commercial Ethernet components, enables remote operation via the corporate LAN or the Internet, and increases operational speed to the LAN choices of 10/100/1000 Mbps of Ethernet. It also exposes the PLC to the usual Internet security problems, requiring the use of protection with known methods. The Schneider divisions have implemented more than 75 products with Modbus/TCP as part of their Transparent Factory initiative. Many other companies have also chosen to use Modbus/TCP as their primary Application Layer for their Ethernet interfaces.

The Modbus application layer commands are given in Table 5-8. The terms used in Table 5-8 were first introduced by Modicon in the creation of relay ladder logic (RLL) for their PLC but are now in common use. The following are the definitions used in RLL:

Coil	A single output bit
Input	A single input bit
Register	A 16-bit assembly of bits or a value
Holding	Internal register storage, typically in the 40,000 range
Force	Set the actual state of an output bit or multiple bits
Preset	Sets a value into a holding register
Mask	XOR (logical exclusive OR) with a mask register before output

Table 5-8. Modbus Command Set

Command	Function	Command	Function
01	Read Coil Status	13	Program Controller
02	Read Input Status	14	Poll Controller
03	Read Holding Registers	15	Force Multiple Coils
04	Read Input Registers	16	Preset Multiple Registers
05	Force Single Coil	17	Report Slave ID
06	Preset Single Register	18	Reserved for Programming
07	Read Exception Status	19	Reset Communication Link
08	Diagnostics	20	Read General Reference
09	Reserved for Programming	21	Write General Reference
10	Poll	22	Mask Write 4X Register
11	Fetch Communication Event Counter	23	Read/Write 4X Registers
12	Fetch Communication Event Log	24	Read FIFO Queue

5.14.1 Modbus Protocol Overview

The primary *Modbus ASCII mode* protocol is designed to pass 16-bit register values across any serial communications line with maximum efficiency and with error checking. Serial communications usually expects a stream of characters encoded using ASCII (ISO-14962-1997). Standard character communications allocates 7 data bits plus 1 parity bit to encode one character. A 16-bit register can be represented as 4 hexadecimal numbers, each reflecting 4 bits of data. Hexadecimal uses the numeric characters 0–9 and the letters A–F to represent all possible combinations of the 4 bits. Each of these characters is then sent on the communications line using ASCII encoding. A colon character begins the message frame and a carriage-return line feed character combination signifies the end of the transmission. Each transmission begins from the Modbus master and is sent to the slave station, which will have an address between 1 and 127 decimal that is placed into the 2-character address location. The Modbus command is sent in the next two characters. Four characters are required to send each data value. At the end of the transmission are the two characters for error checking, called an LRC (Longitudinal Redundancy Check). The LRC for all of the data plus an even parity bit for each character provides the assurance that the received data is the same as the data sent. Figure 5-13 illustrates the data frame for Modbus ASCII mode.

START	ADDRESS	FUNCTION	DATA	LRC CHECK	END
1 CHAR	2 CHARS	2 CHARS	nCHARS	2 CHARS	CR/LF

Figure 5-13. Modbus ASCII Mode Message Frame

To increase the efficiency of Modbus communications, *Modbus RTU mode* was created. A 16-bit register data is again represented as a string of four hexadecimal characters. Two of the hex characters are packed into a single 8-bit symbol and sent on the communications line. This eliminates the ability for the colon character to be used to indicate the start of a message frame and a carriage-return line-feed character combination to be used as an indicator of the end of transmission as in Modbus ASCII mode. The message is framed by at least 28 bits as a silent interval at both the start and the end of transmission. The address and the Modbus command are each sent as a single 8-bit symbol. Error checking may again use the even parity bit and always includes a 16-bit CRC (Cyclic Redundancy Check) value at the end of the transmission to provide assurance that data has been sent/received correctly. Figure 5-14 illustrates the message frame for Modbus RTU mode. Note the time delays for the start and end frames.

Unlike the open Modbus ASCII and RTU modes, *Modbus Plus* is a proprietary protocol but is based upon the same set of Modbus application commands. It functions over shielded

START	ADDRESS	FUNCTION	DATA	CRC CHECK	END
>28 BIT TIMES	8 BITS	8 BITS	n x 8 BITS	16 BITS	>28 BIT TIMES

Figure 5-14. Modbus RTU Mode Message Frame

twisted-pair cable using ANSI/EIA/TIA-485 and ISO/IEC 3309:1991 HDLC (High-level Data Link Control) multidrop protocol.

Modbus/TCP is the most recent incarnation of Modbus in which all data is sent in binary data format. The protocol specification has been assigned to Modbus.org for administration and maintenance. In Modbus/TCP the data stream is encapsulated as a standard TCP/IP data exchange. Modbus has been assigned Internet port number 502 for all of its transactions by the IANA (Internet Assigned Numbers Authority). Modbus commands are sent from the client using a TCP/IP message to the server, which responds with the desired data encapsulated in a TCP/IP data stream. Modbus/TCP specifies the format of both the command and response encoding using Internet standard TCP/IP. Modbus/TCP specifies the details of error detection and recovery for this industrial messaging protocol. By using standard TCP/IP framing, Modbus/TCP allows remote access across the corporate LAN and the Internet, which can be both an advantage and a hazard. LAN and Internet access allow truly remote operation but require that safeguards be used to prevent unauthorized access.

5.15 Profibus-FMS

Profibus began as a layered communications stack, which became known as Profibus-FMS (Factory Message Specification). FMS itself is a subset of ISO 9506 MMS (Manufacturing Message Specification), eliminating those commands not necessary for PLC communications. The Profibus-FMS model is similar to that of Modbus – a network of PLCs communicating with host computers and with each other. Profibus-FMS is a multipeer network allowing communications between PLCs as well as with HMI and other factory devices.

FMS is a very common application layer shared with Foundation Fieldbus and WorldFIP. It is included in the international fieldbus standard, IEC 61158. While the protocol for Profibus is included in the standard, the administration of Profibus is the responsibility of Profibus International (http://www.Profibus.com). The technical work of Profibus is the responsibility of PNO (Profibus Nutzerorganisation eV), which supports technical committees for testing and certification, development of profiles, and system integration.

At present, Profibus-FMS is not being implemented often, having been substantially replaced by Profibus-DP. FMS most often appears as an application layer used with Profibus-DP for older applications. The 9600 bps serial line physical layer and the token-passing data link layer of Profibus-FMS are no longer used except on legacy installations.

5.16 Profibus-DP

Although Profibus was created to be a standard communications link between PLCs and host systems such as HMI, the earlier Profibus-FMS was found to be too slow to support HMI update. When a standard connection with PLC remote termination units (RTUs) or remote multiplexers became a requirement, Profibus-DP was created to solve both problems. The high speed of Profibus-DP, up to 12 Mbps, became its most attractive asset. This makes Profibus-DP both a control level bus and a fieldbus. Profibus International prefers the term Profibus rather than any of its modifiers such as FMS, DP, or PA, but industry continues to use these designations.

Many companies support Profibus communications for their products. Although it began as a German national standard, it has now been fully internationalized. Among the benefits of Profibus is that it is the factory communications standard for Siemens, one of the world's largest system integrators and the

largest manufacturer of PLCs in the world. Integration with
Siemens products becomes much easier when using Profibus.

Rather than using FMS as a programming interface to a network of automation devices, Profibus International has created
an object-oriented method using GSD (Gerätestammdaten or
equipment master data) files. Having the GSD for a device
allows the user to access all of the available data for that device.
There is a GSD for the FMS services. The GSD is very similar to
the DD of both HART and Foundation Fieldbus, but does not
share the same format. These three definition files have been
harmonized by the work of IEC in standard 61804, and also in
ISA104.

5.16.1 Profibus-DP Technical Overview

Although the underlying Profibus protocol has token-passing
functionality, Profibus-DP uses it only for transfer of bus mas-
tership to other bus master devices, such as a backup master.
The primary protocol used for Profibus-DP is master/slave, in
which the bus master station polls all slave devices on a cyclic
schedule for data, ensuring synchronous timing. Profibus-DP
does not use an application layer, but instead all programs
work directly with the data link layer for maximum efficiency
and low overhead. When a slave device receives a cyclic data
exchange request, the slave receives the output data sent by the
master and then sends all buffers of data configured for cyclic
updates.

Data transferred between a Profibus master and the slave is
formatted according to the GSD for the remote device. GSDs
are defined for each device and are part of the certification for a
device to meet Profibus standards. Profibus International has
defined GSDs for many standard devices, but most manufacturers extend GSDs for their own devices.

Profibus-DP operates on shielded twisted-pair cable using
EIA/TIA 485. Speeds can vary with the length of the cable but
are specified from 9600 bps to 12 Mbps. Since there can be only

one master at a time, all communications are half-duplex. The slave stations are multidropped along the cable.

5.17 Profibus-PA

Profibus-PA is a hybrid protocol using Profibus-DP command structures but the same physical layer as Foundation Fieldbus H1. Profibus-PA is intended for use in traditional process control applications where delivery of DC power to the field instrument and support of intrinsic safety are necessary. Unlike Foundation Fieldbus H1, Profibus-PA is a master/slave network that is an extension of Profibus-DP.

Normally, the field instruments are wired to a field junction box where they are terminated in a Profibus DP/PA coupler. Profibus-DP is used as the higher level control level fieldbus to connect PA segments to the control system master. Field instrument power is often supplied from the junction box. Since intrinsic safety barriers do not exist for Profibus-DP, intrinsically safe systems require the junction box with the DP/PA coupler to be located in a safe area and the intrinsic safety barrier is placed on each Profibus-PA segment. IS systems often avoid the use of the Profibus DP/PA coupler, and the Profibus-PA segment is directly wired to an interface board in the controller. A typical PLC can terminate up to four PA segments per interface board.

There are many GSDs developed for Profibus-PA field instruments to make addressing the tuning and setting parameters as easy as possible. These GSDs are defined for specific types of field instrument used for process control by the manufacturer of the instrument and given to the user on computer disk media; they are also placed on the Profibus web site. The GSD is effectively the object directory for Profibus devices. More than 1000 GSDs are listed on the Profibus web site (http://www.Profibus.com) for use in configuring control systems.

5.18 PROFInet

PROFInet is the next step beyond Profibus. Like Foundation Fieldbus, PROFInet is an architecture for a control system that includes a multilevel communications structure. The basic concept is that communications exist between component objects in the network. The PROFInet object model expands the GSD concept of Profibus into full XML (eXtensible Markup Language) based object descriptions. As with all object-based systems, the external or exposed attributes (also called parameters) of each object are the items that can be communicated using well-defined rules. The data format for all PROFInet object attributes is an XML data structure. Configuration of a PROFInet is the action of linking the PROFInet objects through their attributes.

PROFInet uses a series of international standards to define all aspects of its protocol. The underlying technology is based on ISO/IEC 8802-3 standard Ethernet with TCP/IP and the Microsoft COM/DCOM object model and protocol. It is designed to interoperate with other networks but supports Profibus networks fully. The eventual scope of PROFInet is to be used in all operations where Profibus is used today, as long as the physical environment is suitable. Many parts of PROFInet are designed to use commercial Ethernet wiring components or their ruggedized versions now available. While the ability to interface with Profibus networks is fully supported, PROFInet is much more than Profibus on Ethernet.

PROFInet uses a software concept called a *proxy* to model legacy bus devices to make them an integral part of the PROFInet data structure. The proxy is implemented on the fieldbus master device for Profibus systems, although it can be used to interface any other industrial automation network as well. The role of the PROFInet proxy is to transparently make the data of the control system accessible on the fieldbus available to PROFInet. While the proxy software may be added to existing controllers, it is most often implemented with a coprocessor on the Ethernet interface to an existing controller.

PROFInet was created for all types of control systems to make the task of I/O addressing much simpler and more error-free than traditional PLC hardware-oriented addressing. I/O points are named as objects with attributes appropriate for their device type by the end device supporting PROFInet or in a PROFInet proxy. This allows object-oriented engineering tools to create self-documenting programs in IEC 61131-3 compatible languages referring to the I/O by its fully qualified object name.

> ## Fully Qualified Object Names
>
> To most programming languages, an object is a data structure with named attributes. The language usually requires a definition statement telling it that a specific variable name is an object of a certain class definition. The object class definition describes the data structure of each attribute of the object with the name of the attribute. For example: I may identify FT101a as an object of class FlowTransmitter that has been defined with attributes of PV (process variable), HL (high limit), AStat (alarm status), etc. To use the value of the current flow in a program, its fully qualified name would be FT101.PV.

Profibus International makes available a library of PROFInet object classes defined in XML in the same way that it has made text-oriented GSD files available. Most of the GSDs are expected to be made available as PROFInet object classes as well.

5.18.1 PROFInet Technology Overview

PROFInet communications is specified in terms of existing standard communications protocols: Ethernet (ISO/IEC 8802-3), TCP/IP (IETF RFC 2001/1042), XML Ver.1.0 Second Edition (W3C), and COM/DCOM (Microsoft). The *wire protocol* (bits flowing over the wire) is DCOM, with the message encoded in XML and transported via TCP/IP, using an Ethernet wiring plant. Note that there is no mention of Profibus in this protocol stack. Rather, PROFInet itself defines the data structures for the connected objects that are encoded in XML. In fact, most of the PROFInet objects have a direct family relationship to the GSD files of Profibus.

Since PROFInet uses the standard TCP/IP stack, these communications should not be configured for tight real-time activities in which a deterministic response in less than 50 ms is required. For real-time responses between 5 ms and 50 ms, there is a Software Real-Time (SRT) extension for PROFInet that is typically offered by suppliers of PROFInet-based control systems. SRT is a modification of the TCP/IP stack, making it better suited for a local LAN service by removing the adaptations originally inserted for the Internet. Removal of the Internet stack features and the restriction to a short message length of 256 bytes or less make SRT-based PROFInet faster, and deterministic as well.

To further adapt PROFInet for very high speed with response times of approximately 1 ms in applications such as distributed motion control, a hardware assisted protocol called isochronous real-time (IRT) is added to PROFInet. In IRT, the Ethernet hardware at each network device contains a 4-port switch to both allow redundancy, when required, and to eliminate TCP/IP stack overhead, allowing up to 150 axes of concurrent motion control. Only the IRT segment uses these extensions to Ethernet switching technology but remains compatible with standard TCP/IP.

PROFInet includes a user layer designed for implementation using IEEE 61131-3 FBD (function block diagrams) and SFC (sequential function charts), both graphical programming techniques suitable for digital discrete control as well as continuous and batch process control. Power flow or data flow can be connected from each automation object's attributes to form the program logic or computation. The PROFInet web site has a library of "standard" or basic automation objects. Suppliers of field devices and PLCs may also offer more complex objects for use with systems configured with their devices. Figure 5-15 illustrates a logical power flow FBD for an interlock used in discrete automation.

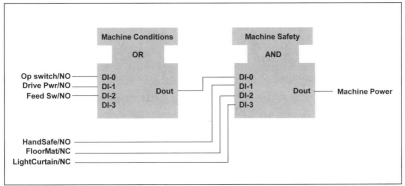

Figure 5-15. Interlock Function Block Diagram

Figure 5-16 illustrates a more complex FBD that might be used for process control.

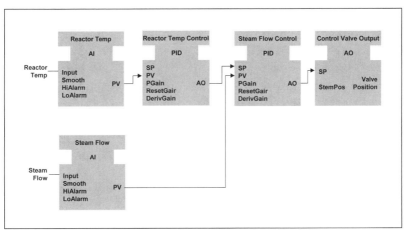

Figure 5-16. Cascade Process Control Function Block Diagram

Figures 5-15 and 5-16 do not necessarily reflect PROFInet or any other product, but do reflect IEC 61499, the international standard for constructing function blocks for FBD. Process control has long used its own version of FBD for configuration of control schemes. When PC-based control began, flowchart programming was introduced. FBD for discrete control is still new, and PROFInet advocates its use.

5.19 Seriplex

Seriplex is included here for completeness. While thousands of Seriplex nodes are still sold annually, they are mostly for installation in plants where it has been used previously. The open specification is no longer supported. Although Seriplex was originally conceived in 1990 as a discrete sensor network, it has been enhanced to include analog values. The original developer's company was purchased by Square-D, which later merged into Schneider Electric. Schneider itself promotes AS-i as its primary sensor network, Profibus-DP as its primary fieldbus, and Modbus/TCP as its primary control bus.

5.19.1 Seriplex Technical Overview

Each device in a Seriplex network contains a Seriplex chip that is assigned a 3-digit address between 001 and 255, using a hand-held set-up terminal. The device is connected to the Seriplex 4-wire shielded cable. The cable supplies DC power to the device and also supplies a separate clock line. The clock line carries a synchronous clock pulse counted by each Seriplex chip. After an initial synch pulse, each chip counts clock ticks and when the count equals the device's address, the device sends its data as single bit discrete or multibit analog. Any configuration of wiring is allowed: star, multidrop, loop, or tree.

While a master PLC is usually used and contains the Seriplex master clock module as well as the DC power supply, these functions are available independent of a PLC. This makes use of Seriplex as a peer-to-peer network with a separate clock and power module. For output points, the PLC, or some other device, outputs the stated value at the time the count reaches the address of the desired output node.

5.20 SDS

SDS (Smart Distributed System) was created by the Micro Switch division of Honeywell about the same time as Allen-Bradley created DeviceNet. In fact, at one time there were plans for a single protocol, but differences in the intended applications and markets prevented merger. SDS was intended originally to be a sensor network, but it has also been expanded into a low-level fieldbus. The specification has long been offered for open license by Honeywell, but no longer is there an independent sponsoring organization. The functions for SDS devices are included in the SDS Component Modeling specification.

Both SDS and DeviceNet use the CAN protocol in their networks, but with quite different application layers. Honeywell has built SDS versions of their complete line of sensors and actuators. In the SDS line, the device may be configured to do considerable amounts of signal processing appropriate for the device. For example, the SDS limit switch has a debounce filter and is capable of counting pulses (switch transitions) or the time duration between transitions. Outputs, such as for solenoid valves, are capable of pulse duration output over a predefined interval, pulse frequency, or a defined pulse count. While full SDS uses the CAN data link layer, it is possible to use the CAN application layer over any other lower layer, including TCP/IP.

5.21 WorldFIP

WorldFIP is the protocol from which ANSI/ISA Fieldbus was developed. Although this work became Type 1 of IEC 61158 in 2000, WorldFIP is also included as originally specified as Type 7 of the IEC fieldbus standard. The original model for World-FIP was as a data acquisition network for the thriving French nuclear power industry. The objective was to collect large volumes of analog and discrete data very rapidly, using minimal electronics at the device, and using only one hole through the nuclear reactor containment wall. Although it has not been used for this purpose, WorldFIP has been used widely in real-time segments of the rest of the electric power industries and the transportation sector in many countries.

The WorldFIP protocol, like Foundation Fieldbus, is based on distributed arbitration, rather than collision detection, master/

slave, or token passing. They both also provide data distribution using publish/subscribe logic, which allows tight time
synchronization for distributed control action.

The versions of WorldFIP are called *profiles* that relate to the
applications. Table 5-9, reproduced from the WorldFIP web
site, describes the WorldFIP profiles. Profile 1 is intended to be
used only for very simple devices, and lacks features supported by higher level profiles. The lowest levels are often
called *MicroFIP or WorldFIP-I/O,* while the higher levels are
often called *FullFIP.*

Table 5-9. WorldFIP Profile Descriptions

Profile Level	Description
Profile 1:	For devices with few configuration options, which start on receipt of a simple command (e.g., basic sensors).
Profile 2:	For configurable devices that handle small amounts of data for configuration and parameter setting (e.g., I/O, more complex sensors). Key data is of cyclic nature. The device may also handle events.
Profile 3:	For devices that handle a lot of data for configuration and parameter setting (e.g., most AC or DC drives). Event handling is essential.
Profile 4:	For devices with total configuration flexibility (e.g., PLCs).

WorldFIP specifies several physical layers for implementation
of different applications. The same low-speed (31,250 bps)
shielded twisted-pair cabling system used by both Foundation
Fieldbus H1 and Profibus-PA is also specified for WorldFIP
when it is used for intrinsically safe, field-powered process
control instrumentation. Additionally, higher speeds (1.0 and
2.5 Mbps) are specified for copper-wired networks. A speed of
5.0 Mbps is specified for fiber-optic networks. All of these specifications are also included in the physical layer of IEC 61158,
the international fieldbus standard.

5.21.1 WorldFIP Protocol Overview

All of the protocol features of WorldFIP are also included in Foundation Fieldbus. WorldFIP introduced the concept of high-efficiency communications using buffer transfer technology. In WorldFIP, this is called MPS (Manufacturing Periodic/aperiodic Services), as opposed to MMS (Manufacturing Messaging Services.) The simplest (Profile 1) WorldFIP field device acquires data on a predefined schedule, or when it changes, stores it into a buffer, and is prepared to publish it on schedule (periodic) or on demand (aperiodic).

Profile 1 is the subset of WorldFIP protocol implemented on the MicroFIP chip and is a proper subset of the FullFIP protocol. While FullFIP offers all WorldFIP services, Profile 1 offers only MPS to move time-critical data with high efficiency. Periodic data communications may be as fast as 5 ms and is typically used to transmit data for fast speed-control loops. Aperiodic data are sent only on request such as responding to a change-of-state alarm. Use of aperiodic transmission is appropriate in systems where change is normally infrequent or slow.

Review Questions, Network Technology, Part 2

1. What is the primary reason to use AS-i?

2. What are the commercial versions of CAN in industrial networks?

3. How would ControlNet be used in an industrial network?

4. What is CIP (Control and Information Protocol)?

5. What is EDS (Electronic Data Sheets) as used for CIP?

6. How can an Ethernet-based network be made deterministic?

7. What are the two forms of Foundation Fieldbus physical layers?

8. What is the similarity between Foundation Fieldbus H1 and Profibus-PA?

9. What is the difference between Foundation Fieldbus H1 and Profibus-PA?

10. What is the unique property of HART?

11. What makes iDA different from other networks used for discrete automation?

12. Describe how Interbus achieves high-speed synchronous operation.

13. What is the difference between LonWorks and DeviceNet?

14. Why is Modbus protocol so popular, even on non-Schneider/Modicon equipment?

15. How does Profibus-DP achieve its high efficiency?

16. How is PROFInet different from Profibus?

Unit 6: Answers for Review Questions

6.1   Review Questions, Industrial Network Basics

1. What are the layers above the ISO/OSI seven-layer communications stack?

   a. User layer and OPC.

2. What is the primary purpose of a sensor network?

   a. Reduce cost of wiring.

3. What are the four topologies used for networks?

   a. Bus, daisychain, ring, and star.

4. What makes a fieldbus network different from a sensor network?

   a. The end-device has programmable intelligence.

5. What is the primary difference between a fieldbus and a control network?

   a. Fieldbus connects intelligent field devices to controllers. Control networks connect controllers to each other and business systems.

6. What is the primary action of a safety bus?

   a. To instantly cause failsafe action on detection of any unrecoverable network failure.

6.2   Review Questions, Network Architecture

1. What are the uses for a user layer protocol?

    a. Isolates the user from the network details. Provides commonly used functionality.

2. Explain determinism. Why is it important for data acquisition and control?

    a. Determinism is defined as the ability to accurately predict the maximum worst-case delay in network access. Control must take action in real-time and not be subject to random delays.

3. What is the purpose of OPC/DX?

    a. The purpose of OPC/DX is to define data independently of both the control system and the data management or presentation system.

4. Why is Microsoft .NET architecture more important for control systems than Java JVM?

    a. .NET provides a platform for OPC while JVM does not.

5. Field Device Tools are used for information access to what level network?

    a. FDT is intended to be used for fieldbus networks.

6. How are WirelessHART and ISA100.11a different?

    a. Although they use the same radio chip and physical layer, they are very different above that layer. ISA100.11a offers many more options to increase performance and lower cost of installation. WirelessHART is much simpler, and may offer enough performance for many applications and to access diagnostic data from legacy HART instruments not having a network connection.

6.3   Review Questions, Sensor Networks

1. Why should you consider use of a sensor network?

    a. To reduce the cost of wiring.

2. When would it not make sense to use a sensor network?

    a. When the I/O points are not clustered.

    b. When the I/O speed requirement is faster than the sensor network can support.

3. What is the most important factor in selecting a particular sensor network?

    a. Which sensor network does my I/O supplier support.

6.4   Review Questions, Discrete Automation Fieldbuses

1. What is the main reason to use a fieldbus?

    a. Reduce the cost of wiring between a remote I/O and the controller.

    b. To bidirectionally communicate with smart I/O devices.

2. Networks used for control must be deterministic. What does deterministic mean?

    a. The maximum worst-case time to obtain data across the fieldbus can be accurately predicted and is not subject to chance.

3. What are some of the advantages and benefits of using smart discrete I/O?

    a. Remove some of the higher speed operations from the PLC.

    b. Reduce the need for high-speed scanning on PLC operation.

c. Remove some of the more repetitive and often forgotten functions from the PLC's logic.

d. A less expensive PLC can be used.

6.5   Review Questions, Process Control Fieldbuses

1. What are the differences between a process control fieldbus and one used for discrete automation?

    a. Process control does not need the high speed of discrete automation.

    b. Process control needs tight time synchronization.

2. What is the difference between distributed control and field control?

    a. Distributed control is always computed in a host controller.

    b. Field control allows seamless construction of a control loop consisting of connected function blocks to perform the closed loop control function with, or without, the participation of a host controller.

3. What are the benefits of field control?

    a. Better control enabling the process to run closer to setpoint.

4. Which fieldbus should be used for field control?

    a. Foundation Fieldbus

6.6   Review Questions, Control Level Networks

1. What are the main purposes of a control level network?

    a. Connect controllers to host systems and HMI.

b. Connect controllers together.

c. Interconnect fieldbus segments.

2. What is the primary technology trend for Control Level Networks?

a. Use of standard Ethernet and Internet protocols.

3. What else distinguishes Control Level Networks from fieldbuses?

a. Control Level Networks are generally higher speed than fieldbuses.

b. No need for intrinsic safety on Control Level Networks.

4. What is the common high level interface to all Control Level Networks?

a. OPC and the Microsoft common object model.

6.7   Review Questions, Control Level Networks, Part 2

1. Which control level networks are suitable for factory automation?

a. Modbus, Modbus/TCP, ControlNet, Profibus-DP, and EtherNet/IP.

2. What is the primary application for Control Level Networks in factory automation?

a. HMI connection.

3. Which control level networks are suitable for process control?

a. Foundation Fieldbus HSE, Profibus-DP, and ControlNet.

4. What are the benefits of using a control level network to supplement a fieldbus?

    a. Reduced cost of home-run wiring.

    b. Interconnection of fieldbus segments.

    c. Fault-tolerance.

5. What distinguishes a Control Level Network from an information technology network?

    a. There is an identical user layer shared with a supported fieldbus network.

6.8   Review Questions, Network Technology

1. What is the best way to choose an automation network?

    a. Study the appropriate networks to help select a control equipment supplier.

    b. Use the network technologies supported by the dominant control equipment supplier for the project.

2. What are the typical environmental problems that influence selecting a network?

    a. Heat, vibration, corrosion, electrical noise, and moving magnetic fields.

3. What role can standards play in making network selection easier?

    a. Standardization should make it easier to integrate a control system with equipment supplied by the "best-in-class" supplier for each component.

6.9   Review Questions, Network Technology, Part 2

1. What is the primary reason to use AS-i?

    a. To reduce the cost of wiring for simple binary 2-state sensors and actuators.

2. What are the commercial versions of CAN in industrial networks?

    a. DeviceNet, SDS, CANopen, CAN Kingdom, SAE J1939.

3. How would ControlNet be used in an industrial network?

    a. As the upper level network to connect subnetworks of DeviceNet or Foundation Fieldbus to a controller.

4. What is CIP (Control and Information Protocol)?

    a. The common user layer for DeviceNet, ControlNet, and EtherNet/IP.

5. What is EDS (Electronic Data Sheets) as used for CIP?

    a. Defines the communications parameters of the device.

6. How can an Ethernet-based network be made deterministic?

    a. Use full-duplex switches and do not share Ethernet segments.

7. What are the two forms of Foundation Fieldbus physical layers?

    a. Foundation Fieldbus H1 and HSE.

8. What is the similarity between Foundation Fieldbus H1 and Profibus-PA?

    a. They use an identical physical layer specification.

b. They both support intrinsic safety and supply power to process control field instrumentation and control valves.

c. They both support signal conditioning and alarming in field instrumentation and control valves.

d. They both support time-critical control in host controllers.

9. What is the difference between Foundation Fieldbus H1 and Profibus-PA?

a. Foundation Fieldbus supports time-critical control in the field device; Profibus-PA does not.

10. What is the unique property of HART?

a. The digital signals are carried on the same wire as a conventional 4-20 mA analog signal.

11. What makes iDA different from other networks used for discrete automation?

a. RTPS (Real-Time Publish/Subscribe) provides synchronous operation suitable for motion control.

12. Describe how Interbus achieves high-speed synchronous operation.

a. Data exists in a slotted ring shift register with exact timing.

13. What is the difference between LonWorks and DeviceNet?

a. DeviceNet uses the CAN chip while LonWorks uses a Neuron chip with higher cost and much more computing capability.

b. LonWorks is capable of performing analog signal conditioning at the chip level that DeviceNet/CAN cannot.

14. Why is Modbus protocol so popular, even on non-Schneider/Modicon equipment?

    a. A license-free open protocol since 1979.

15. How does Profibus-DP achieve its high efficiency?

    a. It uses a master/slave protocol directly on the data link layer.

16. How is PROFInet different from Profibus?

    a. PROFInet is not just a protocol, but also a system architecture.

    b. PROFInet is implemented using TCP/IP over Ethernet.

About the Author

I am now CEO of CMC Associates, Acton, MA, which is to say that I am an independent consultant, and I can give myself any title I want. I have been actively involved in industrial automation work since 1958 when I started doing instrumentation for a small chemical plant of Ethyl Corp. in Baton Rouge, LA. It wasn't too long after I graduated from University of Florida in Chemical Engineering, that I began working on my MS at LSU in Baton Rouge. Paul Murrill and Cecil Smith were in my graduate school Automatic Control class, the first ever taught at LSU. In 1964, I received my MS in Chemical Engineering.

In 1964, I became one of the pioneers in computer control at Union-Camp in Savannah, GA, now part of International Paper. I developed, installed, and operated an IBM 1800 computer for control of both a fast Kraft paper machine and a Kamyr continuous digester, and consulted to our Franklin, VA mill for bleach plant control. To think that the IBM 1800 had less computing capability and disk storage than the very first IBM PC 15 years later. I did all of the software design and Fortran programming for this real-time system that actually performed closed loop feedback advanced control. What a great experience to have been a control systems pioneer.

Foxboro Company was my next stop. I went to work immediately on their PDP-8 based control systems. I led the team to convert their control systems to the PDP-11 as the FOX/2 and 2A. Later I led the team that brought the FOX/1 to market in my first role as a department manager. Next, I became the Marketing guy for the computer control products and planned the successor line, the FOX/1A. My final assignment at Foxboro was in the R&D area where I ran a project to introduce a new architecture into control systems. Along the way, I completed the coursework and obtained my MBA.

With computer control as my specialty, I was recruited by Ken Harple, the founder of ModComp in Ft. Lauderdale, FL, my hometown, as Director of Industry Marketing. ModComp needed control systems software, so I worked with my old friend Dr. Cecil Smith to create a control systems package for their computers. After a financial meltdown at ModComp, I found myself working for Cecil Smith selling his software on ModComp and other computers.

Ken Harple again recruited me for Autech Data Systems, a company he had formed to build process control systems, after he had been forced out of ModComp. This was great fun and gave me the chance to design my own DCS, the DAC-6000, a Faultproof system. It was the first DCS to feature ruggedized 1oo2D fault-tolerant controllers, an Ethernet-based fiber optic network, and a PC-based touchscreen operator console, all exhibited at ISA 1983. In this time-frame, I joined the ISA50 standards committee to help develop Fieldbus. Failure to secure financing caused a complete shutdown of Autech before we could become self-sustaining.

After solving problems for Computer Products, Inc., Analogic, and other companies as an independent consultant, I moved back to Massachusetts to work for Arthur D. Little, Inc., a world-class technology-based consulting company. ADL taught me the mechanisms of consulting. Most of my time was spent in new product innovation, telecommunications, and some industrial automation. One of my projects was to design the mechanism to detect and suppress commercials while recording video broadcasts so that the VCR would fast-forward past the commercials without missing any story material. This innovative project resulted in two US patents: 5,692,093 (1996), 5,455,630 (1995). Licenses were sold by ADL to all VCR manufacturers and it is called Commercial Advance™. I also took over management of the ISA and IEC Fieldbus standards committees.

When ADL began its downward spiral eventually leading to bankruptcy and dissolution, I joined Andy Chatha at ARC Advisory Group. ARC gave me a marvelous platform and the opportunity to influence the automation industry. It was here that I had the chance to spread the word on the use of Ethernet for industrial automation and began this trend toward its widespread use.

Now, in my own consulting company, I have the chance to help many companies, but at a more leisurely pace. Writing books was not my chosen profession, but it is honorable, and certainly fills my days. Recently, I have added publication of the CMC Wireless Report, a bimonthly publication concentrating on the industrial wireless market. Subscriptions to this report support my very active role on the ISA100 standards committee, where I co-chair the User Working Group and the WirelessHART Convergence Subcommittee.

Richard H. Caro, CEO
CMC Associates
2 Beth Circle
Acton, MA 01720-3407 USA
Dick@CMC.us

Index